"This book is a major and much-needed contribution to the climate conversation in Canada. The collective behind it embodies the very politics necessary to win a just transition that is worthy of the name: Indigenous-led, internationalist, rooted in solidarity, and crackling with moral clarity. *The End of This World* advances a holistic, radically reasonable vision of a future worth fighting for—and the authors have tallied the receipts for that glorious moment when the perpetrators of planetary arson get served the bill."

—**AVI LEWIS**, co-author of *The Leap Manifesto*

"As Idle No More, Indigenous peoples, and their allies have repeatedly stated, 'Indigenous sovereignty is climate action.' This book offers a plan to move from our current death economy to a healthier world that centres and validates Indigenous knowledges. Integrating personal narratives, meticulous research, and practical lists, it calls for social movements and all levels of government to work together to address the climate crisis. All that is left is for us to just transition!"

—**DR. ALEX WILSON**, Department of Educational Foundations, University of Saskatchewan

"*The End of This World* makes clear that, in addition to being a crisis of political, economic, and ecological dimensions, Global North-spurred climate change also represents a relational crisis, one in which the bonds between communities—and between humans and other living organisms, including land—have been forcefully and intentionally severed. The authors urge us to refashion these bonds, reminding us that our only chance at avoiding barbarism lies in principled global anti-capitalist and anti-racist solidarity. *The End of This World* highlights the obligation of residents of so-called Canada to oppose, obstruct, and ultimately eliminate extractive industries based here. It also moves us to provide comprehensive reparations for the Global North's centuries-long colonial exploitation, subjugation, and brutal oppression of Indigenous peoples and the Global South."

–**CHUKA EJECKAM,** political writer and researcher

"Environmental apartheid is over. Indigenous land protectors stand at the forefront confronting unbridled destruction. This is not an Indigenous issue, this is an environmental issue, which makes it every human being's problem. *The End of This World* presents an alternative to the scorched earth reality of colonial capitalism: a new economy based upon ecological restoration and the revitalization of human spiritual integrity."

–**DR. DAN RORONHIAKEWEN LONGBOAT,**
Founding Director of the Indigenous Environmental Studies
and Sciences Program, Trent University

# THE END OF THIS WORLD

# THE END OF THIS WORLD

## CLIMATE JUSTICE IN SO-CALLED CANADA

ANGELE ALOOK, EMILY EATON,
DAVID GRAY-DONALD, JOËL LAFOREST,
CRYSTAL LAMEMAN & BRONWEN TUCKER

BETWEEN THE LINES
TORONTO

The End of This World

First published in 2023 by
Between the Lines
401 Richmond Street West, Studio 281
Toronto, Ontario · M5V 3A8 · Canada
1-800-718-7201 · www.btlbooks.com

Library and Archives Canada Cataloguing in Publication
Title: The end of this world : climate justice in so-called Canada / Angele Alook, Emily
    Eaton, David Gray-Donald, Joël Laforest, Crystal Lameman, Bronwen Tucker.
Names: Alook, Angele, author. | Eaton, Emily, 1980- author. | Gray-Donald, David,
    author. | Laforest, Joël, author. | Lameman, Crystal, author. | Tucker, Bronwen,
    author.
Description: Includes bibliographical references and index.
Identifiers: Canadiana (print) 20220227918 | Canadiana (ebook) 20220228027 | ISBN
    9781771136129 (softcover) | ISBN 9781771136136 (EPUB)
Subjects: LCSH: Environmental justice—Canada. | LCSH: Environmentalism—
    Canada. | LCSH: Environmental protection—Canada.
Classification: LCC GE240.C3 A46 2023 | DDC 363.700971—dc23

Cover art by Chief Lady Bird
Text design by DEEVE

Printed in Canada

We acknowledge for their financial support of our publishing activities: the
Government of Canada; the Canada Council for the Arts; and the Government of
Ontario through the Ontario Arts Council, the Ontario Book Publishers Tax Credit
program, and Ontario Creates.

# CONTENTS

# CONTENTS

# INTRODUCTION

In 2009, the oil sands (or tarsands) company Nexen gave funding to the Canadian Defence and Foreign Affairs Institute (CDFAI) to prepare a now-forgotten study called *Resource Industries and Security Issues in Northern Alberta*.[1] The institute hired Tom Flanagan, a conservative academic often called "the man behind Stephen Harper," to write it.[2] Flanagan warned of a possible "apocalyptic scenario" if there were ever prolonged and deep collaboration between environmentalists, First Nations, and the Métis people, among other groups. By "apocalyptic," he meant that industries like oil and gas would have trouble continuing to extract resources and profits in northern Alberta and would no longer be able to flagrantly disregard Indigenous rights. If these groups were to "make common cause and cooperate with each other," Flanagan wrote, they could form "a coordinated movement with the ability to block resource development on a large scale."

A lot has changed since then—the think tank CDFAI has been rebranded as the more benign-sounding Canadian Global Affairs Institute, Nexen no longer exists after being bought out by CNOOC Ltd, and Flanagan has largely fallen out of the public eye—but we think his central point is more relevant than ever. In fact, it's a big part of why we wanted to write this book. Except, from our perspective, "a coordinated movement" between Indigenous peoples, settler environmentalists, organized labour, and many others is the precise opposite of an apocalyptic scenario. We think it's the one thing that could bring us back from our current slide into climate collapse, colonial genocide, and extreme inequality, and towards a better world where we live in balance with land and life.

But this is, of course, much easier said than done. Flanagan predicted that deep and sustained collaboration between groups was unlikely because they wouldn't be able to overcome their different interests and mount a sufficiently large-scale challenge to the fossil fuel and colonial power structures in so-called Canada.* Despite recent inspiring moments of solidarity—from Idle No More, to Quebec and east coast coalitions to stop Energy East and Alton Gas, to cross-country Wet'suwet'en solidarity blockades to stop the Coastal GasLink pipeline—Flanagan has, unfortunately, largely been correct on this point. And since he wrote the report in 2009, the stakes have become so much higher. As we write in the early 2020s, the COVID-19 pandemic has facilitated a growth in wealth estimated at $78 billion for the forty-seven billionaires in Canada while 5.5 million Canadian workers have been thrown out of their jobs, the oil and gas industry is securing plans to expand its production this decade more than any country other than the United States, and chronic underfunding and resource development without consent continue to undermine Indigenous sovereignty.[3]

One reason a sufficiently powerful and coordinated movement hasn't emerged to counter these threats is the targeted efforts from politicians and the oil and gas industry to stop one from emerging. Flanagan himself was actively working to prevent Indigenous solidarity movements, despite dismissing them as unlikely to emerge in the CDFAI report; while he was writing it, he was also working on a book on "how to voluntarily introduce private property rights onto First Nations lands in Canada."[4] Flanagan's proposal, which would extinguish collective Indigenous land rights, found industry backers keen on stopping Indigenous rights from impeding resource development. Meanwhile Harper's Conservative federal government introduced further measures to impede Indigenous and cross-movement resistance to resource extraction, with legislation that criminalized land defence and allowed surveillance of movements by the Royal Canadian Mounted Police (RCMP) and Canadian Security Intelligence Service (CSIS). And in 2012 Harper introduced an omnibus bill that undermined Indigenous land rights and removed protections

---

* Following the decolonial tradition used in various communities, we use the term "so-called Canada" because while the Canadian state claims full sovereignty over the land within its borders, this claim is dubious, unsubstantiated, and much less simple than that. As we will see, Indigenous Nations did not surrender their sovereignty, which brings into question the legitimacy of calling this place "Canada."

for the environment. In the words of Mi'kmaw lawyer Pamela Palmater, resistance was undermined by the government making "conditions so unbearable on reserves that First Nations are forced to leave their communities and give up their lands for resource extraction."[5]

Though later federal Liberal and provincial New Democratic Party (NDP) governments have used much softer-sounding language, their strategies have been largely the same. Successive governments at both levels talk about "reconciliation," "consultation," and "partnership" with the original peoples of these lands, yet First Nations continue to be subject to boil water advisories, Indigenous children's health and education programs continue to be underfunded compared with those of settler children, companies continue to extract on lands that they have no consent to be on, and the Canadian state continues to undermine Indigenous sovereignty at nearly every turn.

But beyond governments' efforts to maintain the status quo, a key reason cross-movement collaboration has been limited is that potential allies have not managed to go much beyond a narrow common cause—saying no to harmful resource development. To transform away from economies built on destruction and death, we need to say yes to much more together. We need a shared vision for a future that is just: where Indigenous and other rights are respected, where everyone has their basic needs met, and where our economies operate in a respectful relationship with nature. In this book, we call this a just transition.

We are a group of six authors who have been working to build this future. Each of us on the author team came to this work in a different way, and we think that together our experiences and the lessons we have learned from our elders and others have taught us what a just future could look like, what stands in the way, and some pathways for how to get there together. This is the vision we hope to share.

Angele Alook grew up in her Nation of Bigstone Cree Nation in northern Alberta, then, like many Indigenous peoples seeking an education, she left to study, eventually completing a PhD in sociology at York University in Toronto. After working as a researcher for the Alberta Union of Provincial Employees (AUPE), she is now back at York as an assistant professor in the School of Gender, Sexuality and Women's Studies. Angele has extensively researched Indigenous school-to-work transitions and Indigenous participation in the labour force of the oil and gas industry in Alberta, including the gendered impacts of the work and the industry. And she has researched how members of her Nation are keeping land-based economies and knowledge alive, including hunting, trapping, berry-picking, and more, even as resource industries severely encroach on their territories.

Emily Eaton grew up in Saskatoon and is now a professor at the University of Regina, where she is one of the few researchers studying the social and political activities and impacts of the province's oil and gas industry from a critical perspective. As the industry has grown to become a significant financial contributor to the Saskatchewan economy, Emily has been looking at how the industry continues to get its way, despite people, including farmers, being aware of its negative impacts on the environment and on human health. Emily has also been involved in efforts to chart a course for rapid decarbonization in the city of Regina and was a co-founder of the Treaty Land Sharing Network, which is working to implement Treaty relationships between rural settlers and Indigenous peoples.

David Gray-Donald shifted from the world of corporate social responsibility to journalism, covering climate justice activism and its struggles with the oil and gas industry, including pipeline struggles like Energy East (which was defeated) and Line 9 (which was not).[*] At the same time, David was coming to terms with his position as someone from a relatively wealthy community and family in Toronto and Montreal, and co-founded the group Resource Movement in 2015, which organizes and mobilizes people with wealth and class privilege to support grassroots social movements. After working as the publisher of *Briarpatch* magazine in Regina for three years, writing about climate justice topics when time allowed, he moved back to his hometown of Toronto in 2020, where at the time of writing, he worked for Between the Lines, the publisher of this book.

Joël Laforest, who grew up in the suburbs of Calgary, was cautiously campaigning for the Alberta NDP in 2015, seeking an opportunity for a break with decades of oil-first conservative rule in the province. But seeing the Alberta NDP's rapid and firm embrace of colonial petro-politics

---

[*] Climate justice is a conceptual framework and a broad social movement that sees climate change not simply as a physical phenomenon, but one that has vastly uneven impacts on different groups of people depending on location, relative wealth, overlapping oppressions such as race and gender, and so on. These factors need to be understood to inform our responses. Climate change is, by its nature, political.

under Premier Rachel Notley made Joël, who was studying labour history and social movements, think deeply about why this was happening and what would need to be in place to break from the all-party consensus supporting extractive industries. Shortly after, he and a small group of like-minded friends launched the *Alberta Advantage* podcast, which has become one of the most popular political podcasts in the country and a thorn in the side of Alberta's right-wing and centre-left politicians alike.

Crystal Lameman had been living in her territory of Treaty 6, unaware of the collective wider movements for Indigenous rights and the environment, until she began making connections with both movements in the summer of 2011. From a brief interaction with a group of students from England's People & Planet, she quickly took an interest in the court case that her Nation, the Beaver Lake Cree Nation, had launched in 2008. The court case, against Canada and Alberta over Treaty violations attributed to the cumulative impacts of industrial development, was the first of its kind to be heard. It became a vehicle for the change Crystal sought, and she built national and international connections to bring attention to the destruction happening on Beaver Lake Cree Nation and its territory. Today, she is deeply committed to supporting the healing of her people and the land, both of which experience the ripple effects of colonization. Crystal's life work and education continue to centre on intergenerational healing.

Bronwen Tucker got involved in community organizing after joining the 2012 Quebec student strikes, studying climate science, and seeing the inequality in our healthcare system as her dad got sick. She started off trying to get McGill to divest its billion-dollar endowment from fossil fuels and joining youth delegations to the United Nations climate talks. Since then, she's worked across different social movements in grassroots organizations and for non-governmental organizations (NGOs) like Council of Canadians and Greenpeace. Bronwen has lived in Edmonton since 2015, where she helped start the grassroots group Climate Justice Edmonton in the wake of many environmental NGOs in Alberta closing down or shying away from confronting oil and gas corporations. She works as a researcher for Oil Change International, which publishes data and analysis on the fossil fuel industry to support the fight for a just transition.

Through our diverse work and experiences at the intersection of Indigenous rights and climate justice, we've noticed that while these movements are getting closer, we do not yet mount the critical threat to business as usual that Flanagan identified. Something holding us back has been that settler-led climate discussion in Canada—including in reports, books, mainstream media, and at environmental NGOs—has been treating Indigenous rights as an afterthought. When Indigenous rights are mentioned at all, they appear as an add-on, in a separate chapter, under their own subheading, or in special reporting focusing specifically on how climate justice relates to Indigenous peoples. In this book, we want to challenge this practice by putting Indigenous rights and sovereignty at the centre of what needs to be done to rescue a habitable planet. We cannot repair our relationship to the environment without also acknowledging and restoring our relationships to one another.

Canada's fossil fuel industry is powerful and organized. It wields tremendous influence over our political, social, and cultural institutions, a theme thoroughly explored in William Carroll's book *Regime of Obstruction: How Corporate Power Blocks Energy Democracy*.[6] But the fact is that Canada's extractive economy only exists in these lands because of a long history of false promises in which the Crown (representatives of the British monarchy, Canada's official head of state) swore Indigenous peoples would maintain their inherent rights and only benefit from the incoming settler societies. From the start, and continuing until today, instead of working within this framework of mutual benefit and respect, so-called Canada has been stealing Indigenous lands and resources and handing them over to fossil fuel corporations to make relatively few people very wealthy.

More and more settlers are coming to realize that an ongoing theft and denial of Indigenous rights is happening here. And this brings up big feelings. Indigenous peoples, who have long known and lived this reality, continue to face systems of colonial control today, despite royal commissions, apologies, and new rounds of promises from successive governments to get things right. And for settlers relatively newly confronted with this reality, the realization can bring a sense of unease and uncertainty. For some, it has prompted a backlash—and an even more fervent assertion of Canada's supremacy and control over Indigenous Nations, peoples, and lands. But it doesn't need to be this way. In this book we acknowledge the ongoing reality of land theft and oppression and offer a vision of how we might begin to undo it. We encourage you to work with us to build a new world, one where Indigenous sovereignty is fully recognized and we live in good relations with each other and the earth.

In brief, we are calling for mass movements, grounded in shared demands for Indigenous sovereignty, that can make big, positive changes happen. Climate action in so-called Canada can't be considered separate from Indigenous rights. In fact, asserting Indigenous sovereignty will require putting limits on the capitalist economy of Canada that has been wreaking so much havoc. Attempting an energy transition without asserting Indigenous rights is simply greening theft—and it is also doomed to fail. Indigenous knowledges and cultures have invaluable lessons for how to live on these lands, knowledges that we need to move from economies of destruction to economies that repair lands and life. We can diminish the power of the fossil fuel industry and move to renewable energies, while reducing inefficient and wasteful uses of energy. We can enjoy comfortable, safe, reliable low-emissions public transit and buildings in both rural and urban areas. Cities and towns can provide great social services like healthcare and education, and can be far less car-dependent. The gaps between rich and poor can be rapidly closed so we can all live better and with a heightened sense of belonging and trust. Far from being unattainable, these changes are already happening, though not fast enough. And joining in social movements pushing for these changes has the added benefit that, instead of being stuck in feelings of despair and isolation, you can participate in hopeful, creative, and engaging communities.

## Principles for a Just Transition

In this book we use "just transition" to refer to a people-powered transition from a fossil fuel–based economy to one that prioritizes economic, racial, and social justice and upholds Indigenous rights. That is to say, it is a transition with climate justice at its core. The term "just transition" originated in the 1970s when the leader of the Oil, Chemical and Atomic Workers Union, Tony Mazzocchi, called for a superfund for workers affected by policies of nuclear disarmament. As the US-based Climate Justice Alliance explains on its website:

> Just Transition strategies were first forged by labor unions and environmental justice groups, rooted in low-income communities of color, who saw the need to phase out the industries that were harming workers, community health and the planet; and at the same time provide just pathways for workers to transition to other jobs. It was

rooted in workers defining a transition away from polluting industries in alliance with fence line and frontline communities.[7]

Years of labour, climate, and Indigenous sovereignty movement efforts have now pushed a just transition to the fore of climate change discussions, but the term is still being applied very narrowly in some contexts. The phrase appeared in the 2015 Paris Agreement and subsequent climate talks, and in recent official government documents and policy in so-called Canada. When the federal government brought in new regulations requiring the phase-out of coal-fired electricity, it established a national task force that visited coal power workers' communities across the country. In 2019 the federal government committed to introducing a Just Transition Act targeted at fossil fuel workers more broadly, but at the time of writing, it is still consulting communities and has yet to release a report or introduce legislation. What is clear is that governments in Canada have mostly treated a just transition as a series of labour market adjustment policies targeted narrowly at fossil fuel workers and their communities.

Luckily, outside of industry and the Canadian government, many others—Indigenous land defenders, union members, youth climate strikers, racial justice leaders, researchers, and more—are putting forth much more expansive frameworks for a just transition that call for transformative changes to an economy built on death so that we can live in good relations with the land and each other. In the Canadian labour movement organizations like Blue Green Canada and the Green Economy Network, as well as national unions, are working within the International Labour Organization's Guidelines for a Just Transition[8] and alongside environmental and civil society organizations to promote an energy transition that prioritizes labour and human rights.

Ideas for a better and decarbonized future also live under a variety of overlapping labels. Global South–led climate justice movements were among the first to solidify these ideas in the context of climate change, most visibly in joint declarations like the 2002 Bali Principles and the 2010 People's Agreement of Cochabamba. In the United States there have been several proposals for Green New Deals or variations like the Red Black and Green New Deal, which centres Black liberation and Indigenous rights. In Canada, The Leap Manifesto in 2015 was an early articulation of these ideas, following closely after the aligned Élan Global declaration in Quebec.* And there are other proposals—including in books like Max Ajl's *A People's Green New Deal*, Seth Klein's *A Good War*, Kate Aronoff and her co-authors' *A Planet to Win*, and others—that generally fit with our vision, as do frameworks for just recovery, abolition, Land Back, degrowth, and ecosocialism.[9] Whatever it's called, if the goal is to live a good life on a living planet for generations to come, we are certain much more has to change than just swapping our cars for electric vehicles and installing solar panels. Our vision for a just transition is decolonial, liberatory, and transformative. It follows the following principles:

1. **A just transition asserts Indigenous sovereignty here and abroad.** That means recognizing Indigenous peoples' inherent rights and sovereignty over ancestral waters, lands, and territories. It requires Canada to stop blocking Indigenous peoples from restoring and reclaiming their land, language, governance, laws, and economies. Canada's mining, military, and other activities outside of its colonial borders must also stop impeding Indigenous sovereignty abroad.

2. **It follows a fair and safe 1.5°C pathway.** Keeping global heating below a 1.5°C increase from pre-industrial levels is the core target in the Paris Agreement, a demand climate-vulnerable communities, like those in small island states, won after years of fighting to stop higher targets that would lock in much more dangerous and likely catastrophic impacts. Reaching this goal while ensuring atmospheric space is available for Global South countries to pursue just development will mean pursuing deep decarbonization that sees Canada reach near-zero emissions well before 2050.[10] This 1.5°C emissions trajectory should be *high probability* and *safe*, which means not buying into fossil fuel industry promises that we can rely on massive amounts of unproven and risky "solutions" like carbon capture, utilization, and storage (CCUS) or geoengineering to kick in at the last minute.

---

* Most Green New Deal proposals emphasize massive new spending to remake low-carbon public and collective infrastructure and services to reach climate targets and improve people's material conditions simultaneously. There are many overlapping visions for a Green New Deal coming out of Korea, the United States, Europe, and elsewhere, some in line with our vision for a decolonial and globally just transition, and others less so.

3. **Polluters and the wealthy pay their fair share.** The central injustice of the climate crisis is that those least responsible are set to be the most affected by it. But it does not need to play out this way. Our vision for a just transition calls for building economic democracy that will allow us to redistribute resources from the small group of corporations and wealthy individuals who have built their wealth on exploitation, carbon pollution, white supremacy, and colonialism to pay for climate solutions and allow a good life for working-class people everywhere.

4. **miyo-pimatisiwin: Meaningful decent work in a caring economy.** miyo-pimatisiwin is a term stemming from nehiyaw (Cree) laws of care, meaning "living a good life."[11] To uphold miyo-pimatisiwin, we must support workers dependent on the fossil fuel economy— whether they are members of labour unions or not—to transition to new, good jobs. It also means recognizing the immense economic value of women's unpaid work, paid care jobs in the public sector like healthcare and human services, and jobs in education that support future generations. But well beyond these measures, we must build a regenerative economy rooted in care that allows opportunities for meaningful work for all and affords everyone access to the land, water, air, food, education, shelter, and community they need to live a good life.

5. **Globally equitable; knows no borders.** A just transition in so-called Canada must not drive the destruction of lands and life elsewhere; instead, we need to carve out space to allow a just transition every- where. After all, Canada is one of the highest emitters of greenhouse gases (GHGs) per person in the world (and that does not take into account emissions in imported products), and the tenth-largest abso- lute emitter, despite having the thirty-ninth-largest population.[12] To live within global ecological limits, we must redistribute, restructure, and reduce our consumption here. For example, we need to build good public transit for all, rather than relying on mass adoption of electric vehicles that would drive "green extractivism" for battery materials both abroad and here at home—where Ontario's "Ring of Fire" mining development is currently being sold as necessary for the batteries needed to tackle global heating. A globally just transition

will also require rich countries like Canada to confront imperialism
and militarism, forgive illegitimate international debts, pay international reparations, undo unfair trade and financial agreements, and
dismantle border securitization so there is freedom to stay and freedom to move.[13]

6. **sihtoskâtowin: Rooted in solidarity.** sihtoskâtowin is a Cree law
that means coming together in mutual support; pulling together
for survival. Land, life, our communities, and the issues threatening
them are deeply interconnected. Recognizing these interlinkages so
that we can work together towards a better world is the best tool we
have to get there. This means a just transition must also transform
inequities based on race, class, gender, immigration status, sexuality,
disability, and other oppressions stemming from capitalism and colonialism. It should follow the principle of "nothing about us without
us," involving and taking leadership from the communities most
affected by an issue in crafting the solutions to it. This solidarity must
also be international, building and strengthening ties with related
struggles in the Global South.

   A detailed vision of a world that upholds all of these principles and a
full roadmap of how to get there would be well beyond the scope of one
book. We focus here on the twin dynamics that we believe have gone
most wrong in so-called Canada to create the conditions we face today:
the violation of Indigenous peoples' inherent rights and sovereignty,
and the fossil fuel economy that relies on this violation. More than
describing the problem, we propose possible paths forward to right
these conditions. We do not think that the vision for a just transition
over the next decade or so that we propose in this book will be enough
to fully dismantle the structures of power that are locking us into a death
economy—namely capitalism, colonialism, and white supremacy. But
there are cracks in many of these forces that are holding up the status
quo, and a just transition in line with these principles would be a massive
springboard in the right direction. Social change is non-linear, and many
new worlds are possible if we stake out a near future where Indigenous
rights are upheld, warming limited to 1.5°C is in reach, and wealth has
been massively redistributed.

## Indigenous peoples, formerly enslaved people, displaced peoples, and settlers

We often refer to Indigenous peoples and to settlers in this book. Indigenous peoples are members of distinct societies with ancestral ties to the land predating colonization; this includes people who identify as First Nations, Métis, or Inuit, whether on reserve, off reserve, rural, and/or urban dwelling.* Originally, the term "settler" referred to Europeans who came and "settled" on Indigenous lands. The term appears in a number of Treaties drafted by the Crown's negotiators. Now "settler" includes a wider group living on Indigenous lands. It is not derogatory, but simply an accurate term. "Settler" as well as "Indigenous" are relational terms rather than racial categories. Being a settler here comes with certain rights and responsibilities.

But not all people who are non-Indigenous to these lands are necessarily settlers. Descendants of formerly enslaved Black people are distinct from settlers, as are people displaced from their homelands elsewhere through colonization and imperialism. Some members of both of these groups are also Indigenous peoples from lands outside of what's now known as Canada. For example thousands of Indigenous peoples displaced from Guatemala by Canadian and US-backed military dictatorships now live in Canada. There is much more brilliant writing on this subject that we encourage you to explore,[†] but for now we emphasize that among Black and other racialized people not Indigenous to these lands, there are a variety of reasons people now live in Canada—while some are relatively class-privileged "economic migrants," others are the descendants of enslaved peoples or displaced peoples. In other words, there is more complexity than just the Indigenous and settler distinction, even if we sometimes simplify things down to that dynamic in this book.

---

\* More and more Indigenous peoples are reclaiming their identities using their Indigenous languages, so when we say Indigenous we include, for example, people who identify as nehiyaw (the Cree word for four-bodied or Cree person), like two of our co-authors. We also cite writers who identify as, for example, Mi'kmaw, Anishinaabe, Tsimshian (Kitsumkalum/Kitselas), and Nuu-chah-nulth (Ahousaht), and we honour their identities by naming the cultural and language backgrounds they use as their identities.

† âpihtawikosisân, "Settling on a Name: Names for Non-Indigenous Canadians," *âpihtawikosisân: Law. Language. Culture*, blog, February 28, 2020, apihtawikosisan.com.

## What We Call Things

Words are powerful. We all know that, and governments do too. In the 1990s, public relations consultants advised the federal government to use "Special Words and Tactics" (SWAT) to deal with Indigenous claims and sovereignty.[14] With this approach, the language of Indigenous rights is adopted—such as "nation-to-nation," "self-government," and "recognition"—but governments assign their own content to these words to maintain colonial relationships and co-opt Indigenous assertions and understandings of these rights and relationships. The exact names and methods have shifted over the years, but words and terms are continually co-opted as a way to dampen demands for Indigenous sovereignty and rights and maintain the fundamental goal of assimilating Indigenous peoples into the Canadian Confederation.

This tactic is used because the terms are popular—or in other words, because they are *powerful*. So in this book, we generally take the approach of claiming these powerful words (like "just transition," "Indigenous rights," "Land Back," "Green New Deal," "decolonization," and others) and contesting the weaker interpretations put forward by governments and corporations. As organizer and social movement theorist Jonathan Smucker put it:

> To challenge entrenched power requires us to claim powerful words and symbols and to make them mean the things we need them to mean. We cannot afford to treat language as static. The meanings of words and symbols are constantly changing, and part of our struggle is to intervene in how their meanings change.[15]

\* \* \*

We believe it is important for just transition discussions to be accessible to all audiences because, fundamentally, what we are talking about in this book is repairing relationships. Repairing relationships with each other and with the land. And key to any relationship is to listen deeply to each other. With that in mind, we invite you to read the following pages with an open heart and mind.

The next chapter begins by situating the discussion of just transition in its proper context: the history, present, and future of Indigenous rights and sovereignty in so-called Canada.

# 1.
# NO MORE BROKEN PROMISES

## Asserting Indigenous Sovereignty

### AS LONG AS OUR CHILDREN LEAD
#### Crystal Lameman[1]

The birds sang me a song, a familiar song, one that my inner child knows, that I, as an adult, remember; it was the welcoming of a new day song. My mind would not sleep throughout the night, tossing and turning, waiting for the new day to arrive. I lay in bed as the birds sang Grandpa Sun a welcoming song, and I wondered if that is why I hadn't been able to sleep. Was my spirit waiting for that familiar song that I missed so much? This event lingered in my mind as I sat in ceremony, close to Mother Earth, surrounded by the ones who love me—those who still walk an earthly life and those who walk with me in spirit. I sat quietly and intently, listening to my elderly relatives talk about this land, in Treaty 6 in northern Alberta. They told me my family has been here for thousands of years and described how we have lived and sustained ourselves. How generations before me gathered and harvested medicines that have kept us alive, from the time the medicine spirits offered themselves as servants and Creator introduced us. Through these teachings I am reminded of how we know this land and this land knows us. This land is where my ancestors' bones feed all that grows around us and nourishes the medicines that we thought we had forgotten—but they remind us that they know and remember us. We were reintroduced today, this first day of summer.

I am reminded about my community responsibilities by Elder

Willie Ermine, knowledge keeper and scholar, as I sit on Beaver Lake Cree Nation, Treaty 6 territory, with the grass beneath my bare feet and the sounds of children playing in the distance, coupled with the singing of the bird kingdom. In this moment I know I am blessed, but I am also painfully aware of how much work is left to be done so that we may again come to that time where the instructions from our Creator are complete and our spirits are full and revitalized. Elder Willie Ermine, reflecting on the process of Indigenous knowledge revitalization and nehiyaw (Cree) process and protocols in nehiyaw knowledge transfer, said, "We have a long way to go before those instructions are complete once again, but this young boy sitting here, he shares the same name as that bird who gave himself to us so he can teach us. This eagle is his relative and they are getting to know one another. This young boy sits here taking care of the smudge, and in doing so he gives us hope." The presence of my young son, who took the seat of the ceremony helper, brought me to a place of knowing that as long as our children lead where we lead, walk where we walk, teaching and guiding us, we are on our way to reaching miyo-pimatisiwin (a good life).

My understanding of wellness as it relates to land is a lived experience, as I refer to nohtâwiypan, my late father, who spent time in the bush and gave himself back to the land as a way of being well again. I can recall conversations with him in the last few years of his earthly walk in which he spoke about a time in his life when he was in his best state of mind, when he was on the trapline and living in the cabin on the "island" with his father.* I did not have the same experience nohtâwiypan had with his father, which means that I did not fish, hunt, or gather with my father. However, after nohkômpan, my late paternal grandmother, took me from my parents, I grew up gathering with her; and my children were able to hunt, fish, and gather with my late father, their mosom (grandpa). These affirmations of land-based learning and cultural practices remind me that the bones of those I love, my ancestors, are in the roots, the medicine, the air that we breathe and the minerals in the ground. The

---

* The "island" is a culturally significant area in the Beaver Lake Cree Nation. It was once surrounded by water and the trapping cabin that has been there for over a hundred years was used by every trapping family in the Nation. Beaver Lake has receded so much that the "island" is no longer.

exercise of traditional land-based practices and Indigenous knowledge of interconnectedness play a role in breaking down the continued attempts at genocide and the transmission of intergenerational trauma. My little family is living proof of that.

My story contains within it the seeds for understanding what holds us back from a just transition, and for how to overcome those obstacles so that we can get there. Relationships have been damaged, but there is a movement of people actively engaged in righting them.

## Beaver Lake Cree Nation's Treaty Case

In 2008 the people of the Beaver Lake Cree Nation, co-author Crystal Lameman's Nation, who live in an area of boreal forest the size of Switzerland in so-called Alberta, filed a lawsuit against the Crown for infringements to their Treaty as it pertained to overdevelopment, industrialization, and the cumulative impacts of this development.[†] They did this to protect their land for the generations to come. This land had been home to large herds of mammals like moose, caribou, and elk. Hundreds of freshwater lakes and rivers provided clean water and an abundance of healthy fish. But now, years of activity from the oil and gas industry and other industrial development have poisoned the water, eliminated whole forests, and decimated traditional food sources for the people of Beaver Lake Cree Nation.

Beaver Lake Cree Nation was the first to ever challenge and be granted a trial on the cumulative impacts of industrial development as a Treaty right infringement. Since the claim was filed, two other First Nations (Carry the Kettle and Blueberry River) have filed similar claims.[2] In 2021, the British Columbia Supreme Court ruled that the Treaty 8 rights of Blueberry River First Nations have been breached by development authorized by the province. Blueberry River First Nations will receive funds to restore and heal the land and protect Indigenous ways of life, and the province will enter into negotiations on the long-term protection of Treaty 8 rights.[3] The agreement does not halt all development, with several projects proceeding and others cancelled until further negotiations. Given the accelerated effects of development

---

† Canada is a constitutional monarchy—in practice the "Crown" in Canada is the Queen of England, the House of Commons, the Senate, and the thirteen provincial and territorial legislatures.

across Treaty territories, more claims about the cumulative impacts of industrial development can be expected.

In the past, Canadian courts had been limited to cases that consider the impacts of one project or mine at a time. Beaver Lake Cree Nation's case was the first to consider all of the projects, collectively. Alberta and Canada have far exceeded the land's capacity for development by taking up more than 88 percent of the Beaver Lake Cree Nation's traditional territory for industrial development, such as oil and gas wells. They have recklessly authorized oil sands, fossil gas (also known as "natural gas"), and forestry projects, military facilities, and other development without any real regard for the rights of Beaver Lake Cree Nation and other Treaty Nations. As a result, the Beaver Lake Cree Nation territory is now covered with over 35,000 oil and gas sites, 21,700 kilometres of seismic lines, 4,028 kilometres of pipeline, and 948 kilometres of road. While any one of these projects by itself might be tolerable, when taken together they threaten to destroy the way of life and the land that has sustained the people for thousands of years. Industrial development like oil sands extraction has polluted most of the water bodies in the region. The sheer size of the area occupied by oil and gas wells and other infrastructure has displaced moose and elk and decimated the caribou population, and most traditional hunting and gathering grounds are no longer available or accessible.

In 1876, the Crown and the ancestors of Beaver Lake Cree Nation entered Treaty 6, whereby the Crown made a solemn promise that Beaver Lake Cree Nation would be able to maintain their way of life. The Crown, by way of the federal and provincial governments, is failing to uphold Indigenous peoples' inherent and Treaty rights such as fishing, hunting, trapping, and gathering plants and medicines. The case serves as a vital precedent for Treaty Nations, and indeed for everyone in Canada. If industrial development can be limited in order to maintain the healthy ecosystems that are essential for Treaty rights, vast swaths of land, air, and water can be protected for future generations. After being delayed for years while Canada and Alberta tried to get the case thrown out, the Beaver Lake Cree Nation case is scheduled to go to trial in the Alberta Court of Queen's Bench in 2024, sixteen years after the Nation began legal proceedings.

But in the meantime, the Nation is facing an infuriating battle over advance costs. During the height of the COVID-19 pandemic, Alberta and Canada chose to challenge the Beaver Lake Cree Nation's successful application for advance costs to assist in legal expenses. In June 2020, Alberta's Court of Appeal agreed that the Beaver Lake Cree Nation government should have to exhaust all of its resources before the Crown would provide costs to pursue the case.[4] In the court proceedings, Beaver Lake Cree Nation had to provide proof that they were "poor enough" to require the Crown to contribute to the advance costs of pursuing the case, when the small amount of money that the Nation has in reserve is for things like repairing the water truck, which brings clean water into the Nation, where over 85 percent of homes are not connected to the main water line. This money is for emergencies, like when the school flooded a few months after the advance costs were overturned. This court decision was dehumanizing and further perpetuated violence against Indigenous peoples.

Yet again, Beaver Lake Cree Nation had to divert more funds and resources to apply for leave to the Supreme Court of Canada. The decision from the Supreme Court came down in March 2022, and it verified that a First Nation should not have to completely exhaust its funds to assert its inherent and Treaty rights in the colonial court system, since having funds in reserve is a requirement to functioning as a Nation and government. However, the Supreme Court did not completely overturn the appeal court decision; it sent the Beaver Lake Cree Nation advance cost case back to the lower court to rule again on the three-part advance cost legal test. In the case of *BC v. Okanagan Indian Band*, the test was determined based on demonstration of impecuniosity (poverty), merit of the case, and issues of public importance.[5] In the Beaver Lake Cree Nation case, the Supreme Court went on further to state that "pressing needs are not defined by the bare necessities of life. Rather, and in keeping with the imperative of reconciliation, they ought to be understood from the perspective of that First Nation government."[6] The Supreme Court recognized that pressing needs include basic necessities of life, including adequate housing, a safe water supply, and basic health and education services, and may also include spending to provide enhanced health and education services or to promote cultural survival. The Supreme Court also stated that it was open to courts to take judicial notice of the systemic and background factors affecting Indigenous peoples in Canadian society, which include the history of colonialism, displacement, and residential schools—insofar as they may be relevant to understanding a First Nation government's financial situation and spending priorities.

This decision in *BC v. Okanagan Indian Band* provides the Beaver Lake Cree Nation with solid ground to go back to the lower court to argue for advance costs, but the Nation will have to endure yet another colonial court proceeding. If in fact the Crown was serious about reconciliation, it would not drag Beaver Lake through another court proceeding and instead negotiate a resolution. To date, the Crown has refused to negotiate any costs for the litigation, nor has it addressed the issues raised in the Nation's underlying litigation. The continued red tape, roadblocks, and refusal to negotiate are the kinds of intentional efforts used to silence us as Indigenous peoples.

## Colonial Impositions

The Beaver Lake Cree Nation case is one example among many of how Indigenous peoples, asserting their inherent role as stewards of the land, stand in contrast to the typical approach of resource companies and the colonial state. This stewardship work is vital, as many of the last remaining forests and much of the planet's biodiversity, lakes, rivers, and streams are in Indigenous peoples' territories—not just in so-called Canada but around the world. Indigenous peoples have been sustainably managing natural resources through customary laws and sustainable resource management practices for thousands of years. Indigenous peoples are the first environmentalists. The protection and promotion of these traditional systems is crucial in the fight against climate change, deforestation, and water pollution. Yet it is infringed on by rampant large-scale development projects.

Indigenous peoples make up 6.2 percent of the world's population and account for 19 percent of the world's poor, yet they are responsible for maintaining 80 percent of the biodiversity of the planet and they steward 22 percent of the world's surface.[7] This is an incredibly high amount of biodiversity compared to the proportion of land, which shows that Indigenous peoples have been stewarding the lands and waters with care. Many of these traditional territories are also carbon sinks, helping mitigate global heating, and have been protected for generations by Indigenous peoples.[8] Indigenous peoples are asserting their land rights to stop many of the most harmful destructive projects—blocking fossil fuel projects in the 2010s equivalent to at least 1.5 gigatons of $CO_2$ in what's currently called the United States and Canada.[9] Indigenous peoples have maintained invaluable knowledge connected to culture, cosmology, and worldview that is becoming undeniably critical to building solutions, as well as mitigation and adaptation strategies, to climate crises around the world.

But this knowledge is not being respected and uplifted; it boils down, in much of the world and specifically in Canada, to colonial impositions on Indigenous lands and into Indigenous ways of life, and to broken relationships. We need to revisit the existing relationships we have together, because it is there that we can find the paths forward.

## Treaty Relationships

"What I have offered does not take away your living; you will have it then as you have now, and what I offer now is put on top."

—Treaty Negotiator Alexander Morris during
Treaty 6 negotiations, 1876[10]

There may not be a shared understanding of what rights Indigenous peoples have. The most fundamental rights are inherent rights, rights Indigenous peoples are born to—the right to air, water, land, natural resources, governance, to spiritual ways of knowing, being, and doing. As co-authors Crystal Lameman and Angele Alook can attest, these rights have been practised since our Creator placed us here on this land and well before the arrival of Europeans. They are collective rights that flow from our bloodlines and continued relationships to our home territories, and are distinct to the practices of each Indigenous Nation. Inherent rights are not granted by an external authority—such as a colonial government like Canada or the United Nations—and cannot be taken away, because they are the very rights we are born with.

There is some confusion about this, in large part due to misunderstandings about Treaties, misunderstandings long perpetuated by Canadian and provincial governments. So it's important to understand the Treaties. The practice of Treaty making was not introduced when Europeans arrived; rather, there was a long tradition of Treaty making between Indigenous Nations, and between Indigenous peoples and their non-human relations in their territories. An example is the Ojibwe (Anishinaabe) story of The Woman Who Married a Beaver. In this story a young girl is transformed into a beaver; she marries a beaver and must maintain relations with humans. It is a teaching of transformation, love, sharing of resources, trust, mutual respect, and treating each other in good ways. In this story, Anishinaabe people make gift offerings to the beavers, return the bones of beavers to the water, and in return the beavers give themselves as food back to the people.[11] All the while, the humans do not speak ill of the beavers and do not do anything that may harm the beavers, because they must live in mutual respect and coexistence with them. To Indigenous peoples, writes Cree and Saulteaux scholar Gina Starblanket, "treaties are sacred undertakings" based on localized Indigenous knowledge systems.[12]

When Europeans arrived, there were many processes of Treaty making with Indigenous Nations. In the Maritimes in the 1700s, many Peace and Friendship Treaties were made, and they are still in place today. As the name suggests, these Treaties were to facilitate peace as well as trade. They were not land surrenders. In the Great Lakes, St. Lawrence, and Ottawa River regions, a number of Wampum Belt Treaty agreements were made between settlers and Indigenous Nations including the Haudenosaunee (Iroquois) and Anishinaabe. These are real belts made of wampum—beads made from shells—that these Nations use to mark agreements between peoples. Important belts include the Two Row Wampum, which describes two boats going along a river together in perpetuity, neither interfering with the navigation of the other's vessel. And in 1764, a Covenant Chain Wampum belt was exchanged, part of the Treaty of Niagara and 1763 Royal Proclamation process that set the foundation for the establishment of Canada in much of what is now Ontario and Quebec. These, like the Peace and Friendship Treaties, were not surrenders or land transfers, but agreements between Indigenous and settler peoples to live together and share the land.

The same can be said about the post-Confederation numbered Treaties, Treaty 1 to Treaty 11, agreed to from 1871 to 1921. These Treaties spanned from northern Ontario, across the Prairies, and into the northwest. For example, the Beaver Lake Cree Nation, to which Crystal's family belongs, entered into Treaty 6 in 1876. Angele's people, in Bigstone Cree Nation, entered into Treaty 8 in 1899. Both the Beaver Lake Cree Nation and Bigstone Cree Nation are located in what is also known as northern Alberta. But we want to be very clear. When First Nations entered Treaty, they did not give up their inherent rights. In fact,

First Nations were assured that the Treaties affirmed their right to self-determination and to live and exist as sovereign peoples on their lands and with their resources. Furthermore, First Nations were promised they would live as they did the day before Treaty making, and that what was offered in Treaty making was additional benefits. First Nations never interpreted the making and entering of Treaty as leasing, ceding, or surrendering their lands, territories, and/or resources. Through the making of Treaty, the Crown made many promises to our peoples—for example in Treaty 6 on the Prairies, the medicine chest and famine and pestilence clauses—in exchange for the use of our lands to the "depth of the plow" only. These promises the Crown made are what are now called "Treaty rights." Minerals were never put on the table for discussion; therefore, Indigenous peoples never relinquished rights to their water or minerals, nor to the Rocky Mountains.*

Alexander Morris was the Treaty Commissioner who negotiated a number of Treaties on the Prairies, including Treaty 6. He was also the Lieutenant Governor of the North-West Territories, which was the name Canada used at the time to refer to a vast area of what is now called Western Canada. Morris, in the negotiations leading up to the 1876 Treaty 6 agreement, promised that First Nations could continue to hunt, fish, trap, and gather, while continuing our land-based way of life into an indefinite future. No end point was discussed, which is clear in the Treaty text: "As long as the sun shines, the grasses grow and the water flows."

The spirit and intent of the Treaty was made clear by the Crown's Treaty Commissioners themselves in the Treaty-making process. As Commissioner Morris said in 1876, "what I offer now is put on top" of your existing way of life. Through Treaty, then, First Nations added new, negotiated rights on top of their inherent rights.

---

* Until quite recently the Alberta Rocky Mountains were protected from open pit mining under the 1976 Coal Development Policy. In 2020 under the United Conservative Party led by Jason Kenney, the Alberta government announced plans for open pit mining on the eastern slopes, which would be a threat to the headwaters of major rivers. The Kenney government rescinded the Coal Development Policy without public consultation or input from First Nations or the scientific community; this egregious action resulted in public outcry against the government, with notable Alberta celebrities and First Nations communities in Treaty 7 territory speaking out publicly against this development. For more information please see: protectalbertawater.ca.

### First Nations' Inherent and Treaty Rights

By the end of the Treaty-making process, Indigenous peoples had confirmed, at least, the following collective rights:

1. We have sovereignty
   a. Our peoples can act within their own authority as derived from the Creator
   b. We have complete jurisdiction over all matters that pertain to our peoples, lands, and territories
   c. We can enter Treaty with other Nations
   d. There are collective rights of Nations
   e. We have internationally recognized sovereignty
2. Right to land, air, and water
   a. Indigenous peoples own and occupy our lands and territories including all natural resources
   b. We possess historical continuity to our lands and territories since time immemorial
   c. Our peoples have the obligation to protect the land and live in harmony with the earth
   d. Indigenous peoples are keepers of the lands for our unborn generations
3. We have spiritual beliefs, customs, and traditions
4. We have our own languages
5. Self-determination
   a. Creator's laws (natural laws)
   b. Indigenous laws and customs
   c. Institutions of governance
6. Justice systems
   a. Indigenous law enforcement

Indigenous knowledge keepers hold the understanding of Indigenous peoples' rights and pass it along, and the rights are lived out in our Nations. But colonial governments have tried to undermine these rights since the time of Treaty making. As Starblanket warns, "The [European] transactional understanding [of Treaty] suggests that Indigenous populations ceded and surrendered title to the land to the Crown in exchange for a fixed spectrum of rights and entitlements." For Canada, this colonial mythology about Treaty "facilitated both settlement ('the opening up of the North') and capitalist pursuits ('access to valuable natural resources')." This interpretation of Treaty provided a narrative of the "peaceful absorption of Indigenous peoples into the fledgling Canadian state."[13]

This issue of whether Indigenous peoples gave up their rights to the land is one of the most highly contested issues of Treaty interpretation, at least from the settler perspective. But as the report of the Canadian government's 1996 Royal Commission on Aboriginal Peoples explains, "it is highly probable that no consent was ever given by [Indigenous] parties to that result. [Indigenous] people, who believe the Creator set them on their traditional territories and gave them the responsibility of stewardship of the land and of everything on it, are not likely to have surrendered that land knowingly and willingly to strangers."[14] Indigenous peoples did not believe one could own the land; therefore, the sale of land did not make sense. To Indigenous peoples, Treaties were regarded as land-use frameworks and agreements for sharing the land in mutual coexistence.

Canada and the provinces of the numbered Treaties that span from northern Ontario to the Yukon, by virtue of the Imperial Crown, did not receive anything but instructions on how to share and live peacefully with the original peoples of this land. Yet Canada and the provinces claim to own and have jurisdiction over nearly everything in and on these lands. Canada's deceitful interpretation of Treaty led to the violent displacement of Indigenous peoples from the lands, into an impoverishing Indian reserve system that made them wards of the state, and refused to recognize Indigenous sovereignty and the responsibilities of Indigenous peoples to land stewardship.

### Restoring Indigenous Sovereignty and Nationhood

In 1876, the same year that Treaty 6 was entered into, the Canadian government also brought into force the Indian Act. While the Treaty Commissioner was entering sacred agreements with First Nations that promised recognition and exercise of inherent rights to self-determination, livelihood, and much more, Canada was imposing new colonial laws and legislation meant to control Indigenous peoples. This racist Indian policy established the colonial framework that undermined the Treaty relationship—despite the fact that section 91(24) of the 1867 British North America Act, which formed the legal basis of Canada, in naming "the exclusive Legislative Authority of the Parliament of Canada" over "Indians, and Lands reserved for Indians," affirmed the federal government had the responsibility under Confederation to address the promises of Treaty. Instead, the Canadian government simultaneously broke its promises by imposing oppressive laws regulated through the sexist and racist Indian Act, which confined Indigenous peoples to Indian reservations through a pass and permit system, outlawed traditional ceremonies, and laid the legal foundation for a system of apartheid and genocide.

Among Indigenous peoples, we sometimes refer to this as an ongoing apocalypse. For example, there was ongoing famine and disease that we were not assisted in treating, ignoring the famine and pestilence clause in our Treaty. Then there was the taking of our children through Indian residential institutions, the Sixties Scoop, and the ongoing discrimination of the child welfare system. And the decimation of the buffalo and subsequent other species, like the caribou that Crystal's people subsist on, which fall under the exercise and practice of inherent and Treaty rights. The experiences of genocide do not stop there, with the militarization against Indigenous peoples, police brutality, access to basic human rights like clean and safe drinking water, the disproportionate rate of Missing and Murdered Indigenous Women and Girls as well as Two-Spirit people, men, and boys.[15] This long list of acts of devastation and destruction is the foundation on which the relationship with the original peoples of this land is built.

"Settler colonialism arose when European colonists did not leave or relinquish power and instead continued to occupy Indigenous lands, setting up Euro-Canadian political and economic institutions," writes Cliff Atleo, a Tsimshian (Kitsumkalum/Kitselas) and Nuu-chah-nulth (Ahousaht) scholar. Atleo also notes how Indigenous scholars "stress that colonialism in Canada is not simply a legacy but a persistent reality that Indigenous peoples still endure."[16] In the face of this, Navajo scholars Andrew Curley and Majerle Lister write:

> For generations, Indigenous peoples were able to survive on their lands through strategic engagement with extractive industries and capitalism. The legacies of these practices scar the landscape. They helped us survive on the land but also destroyed much of it in the process. With colonization, Indigenous peoples saw their lands taken and lives permanently altered. This constituted its own dystopia.[17]

Some of this history, the current reality, and the effects on our Nations are finally becoming somewhat better known in settler society, especially regarding the violent legacy of Indian residential institutions. Less widely known are our traditional and still-existing knowledge and governance systems. In nehiyaw culture, askiy oma (this land) is governed through nehiyaw ways of balance, reciprocity, sharing, and caring, where we only take from our Mother Earth what we need. And this governance is built on good relations with all land, air, and water relatives. At one point we lived in balance through miyo-wîcihtowin (the act of having good relationships). We understood that our future as a people relied on our children, our intergenerational relations within our Nations; we understood that living well meant respecting the connection we had with life around us.

Good relations meant good, equal autonomous gender relations and healthy family relations. Indigenous women were highly regarded in nehiyaw and many other Indigenous societies for being life-givers (because of their ability to bear life they were regarded as spiritually powerful). And for their crucial roles in the collective good, they were political leaders. Indigenous women held economic power in the distribution of resources. For example, among the nehiyaw, nehiyaw-iskwewak (Cree women) were considered to have authority over the home and all of its contents, and women decided how to divide the meat among the family when there was a big hunt. Prior to colonization, older women and clan mothers were important decision makers, especially regarding land issues. For example, among the Haudenosaunee "female-led clans held the collective land base for all the nations of the confederacy."[18] Additionally, as Alex Wilson, a Two-Spirit member of the Opaskwayak Cree Nation writes, nehiyaw societies respected the identities and roles of non-binary people.[19] Two-Spirit community members were understood to carry both male and female energy, and they brought balance to the community by carrying out various roles.[20]

We have continued these ways, but they have been under relentless attack. As part of the colonial drive to undermine and dispossess us of our political, social, and cultural systems, our traditional gender relations came under attack. Specifically, the Indian Act replaced our traditional leadership structures where women were important leaders in their own communities. Instead, patriarchy—the subjugation and oppression of women and gender non-conforming people by men—was inserted into

our communities. Canadian governments and citizens established economies and governance systems on our lands that are patriarchal and also
capitalist, based largely on extraction of resources with profits going to a
relatively small number of people. The colonial project tried to eliminate
Indigenous peoples as rights-holders in order to extract wealth from the
land and settle newcomers with new economies.

We believe Kanienkehaka scholar Audra Simpson is right when
she says that "Canada requires the death and so called 'disappearance'
of Indigenous women in order to secure its sovereignty."[21] Indigenous
women play a central role in our communities, holding essential knowledge and relations. The state, Simpson writes, works through "a death
drive to eliminate, contain, hide, and in other ways 'disappear' what fundamentally challenges its legitimacy: Indigenous political orders."[22] This
violent assault continues today in many ways. The 2019 National Inquiry
into Missing and Murdered Indigenous Women and Girls (MMIWG),
for example, reported that its findings "tell the story—or, more accurately, thousands of stories—of acts of genocide against First Nations,
Inuit and Métis women, girls, and 2SLGBTQQIA [Two-Spirit, lesbian,
gay, bisexual, transgender, queer, questioning, intersex, and asexual]
people."[23]

This is not the relationship we as Indigenous peoples agreed to in
Treaty. Today, we remain steadfast in ensuring that Canada lives up to
the sacred agreements that it made with our Nations, as we have been
living up to our promises to share the land in peaceful coexistence. The
first step to decolonization and reconciliation must then begin with hon-
ouring the original agreements that allowed for the existence of settlers
in these lands. The only way to address systemic racism is for Canada to
recognize the full implementation of our sovereignty, and for Indigenous
peoples to revitalize our original governance systems created by and for
our Nations, grounded in an Indigenous worldview. Without Canada's
recognition and full, unqualified implementation of Indigenous self-
determination, and without our Nations exercising and practising our
own laws, we risk being subsumed by the Canadian legal framework and
absorbed into the Canadian body politic. Canadian and provincial governments keep trying to push nice-sounding frameworks on us, using
our own words, like "self-determination," but these are again colonial
structures that deny our sovereignty and nationhood, and do not offer
decolonization. The path forward must be grounded in recognition and

## How the Canadian State Benefits from Stolen Lands

- Canada leads in mining extraction globally, producing over 60 minerals and metals, with total domestic exports valued at $81.4 billion per year. This amounts to over 20 percent of total national exports.
- The forest sector generated $24.6 billion to Canada's GDP in a single year, $24.2 billion to Canada's trade balance, and 186,838 jobs in 2017, not counting all the secondary industries to which it contributes.
- Fossil fuel production is high, comprised by crude oil (41.4 percent), natural gas (36.5 percent), and coal (9.2 percent),where by far the largest supply of crude comes from the Alberta tarsands.
- Pulp and paper in 2010 pumped $9.8 billion in exports from newsprint and wood pulp into the Canadian economy.
- Canada consistently ranks near the top of global newsprint and wood pulp production that derives from boreal forests.
- It is not only the land that is being exploited. Natural resource sectors were responsible for around 84 percent of total water consumption in Canada in 2005.

Source (verbatim): Hayden King, Shiri Pasternak, and Riley Yesno, *Land Back: A Yellowhead Institute Red Paper*, Yellowhead Institute, October 2019, 25–26, yellowheadinstitute.org. The source for each of these stats is specified in the report.

full implementation of our inherent and Treaty-based models and frameworks, and funding arrangements based on the original Treaty promises.

## Neo-colonial Resource Extraction on Indigenous Lands

Resource industries, touted as drivers of prosperity in Canada, are based on encroachment onto Indigenous lands and infringements of inherent and Treaty rights. As the *Land Back* report by the Yellowhead Institute puts it, "Crown Lands have long catalyzed economic growth for this country."[24] Crown lands—lands that the federal or provincial governments claim to own and control, as opposed to private property or First Nations reserve lands—make make up the vast majority of the land in Canada. Simply put, Crown land means corporate access to natural resources, and this access has made Canada a resource-rich nation.

Angele interviewed members of her First Nation in the heart of

Treaty 8 territory, which is in the middle of the Athabasca oil sands deposit, as part of the Just Powers: Energy Transition and Social Justice research project. Interviewees were all too aware of the land theft and the socio-economic inequality created by this neo-colonial system. First Nations Lands Manager Troy Stuart noted that we are sitting on billions of dollars' worth of natural resources, and each week we watch this wealth leave our Nation in the form of trucks hauling away the resources, and in the form of settler workers who come to work in the oil patch and are driven in and out of our community by the busload; we watch our natural resources and the incomes that are earned from these resources drive past us.[25] According to Stuart, a traditional hunter and fluent nehiyawewin (Cree language speaker), what is propping up the Canadian economy is underneath our feet and we are not able to fully benefit from this resource, as we suffer from housing shortages and inadequate education and health funding. In Bigstone Cree Nation, this inequity has been even more stark during the pandemic, as many mourn our elders and most vulnerable dying of COVID and see youth lost to suicide. And, as has long been the case through oil price booms and busts, we see Indigenous workers among the last hired and first fired in the oil sands. Across the country, during the pandemic, Indigenous and Black youth have suffered among the highest job losses in the economy.[26]

Melina Laboucan-Massimo, a Lubicon Cree climate justice warrior and Indigenous sovereignty and women's rights activist, writes of the environmental, social, and health impacts of fossil fuels in her home in Treaty 8 territory:

> The Alberta tar sands are scarring the earth—polluting and draining
> watersheds, poisoning the air and destroying the land I call home. The
> landscape is drastically changing from a once pristine and beautiful
> boreal forest to an increasingly industrial and toxic terrain. Animals
> and fish have become sick with tumours, and caribou are now listed as
> an endangered species. People are no longer safe to harvest traditional
> medicines, teas or berries because they have become contaminated—
> and even though we fear that our medicines have turned into poison,
> we continue to forage (and forge) the path ahead. People young and
> old have started to die of rare forms of cancers that we have never seen
> before. I come from a community where, until my generation, my
> family was able to live sustainably off the land.[27]

Laboucan-Massimo notes this destruction is ongoing and widespread. "The tar sands are not an isolated incident," she continues; "neo-colonialism in the form of resource extraction is happening across Turtle Island [North America] and throughout Mother Earth. Today the earth is being contaminated and destroyed at an unparalleled rate, and people and animals alike are being sacrificed for the benefit of the greedy few."

## Indigenous Nations Lead the Way

As colonial governments are developing climate change policies, Indigenous peoples should be actively engaged in the research, recommendations, and development of the suite of policies that have potential to both negatively and positively impact their rights and title. When it comes to discussions on developing progressive climate policies, Indigenous peoples' rights are not currently being fully addressed in any of the major oil and gas producing regions of Canada (British Columbia, Alberta, Saskatchewan, Nova Scotia, Newfoundland and Labrador) or internationally. Indigenous peoples have asserted and fought for their rights within climate discussions, yet have not been given decision-making roles within the policy development processes. The solutions that are developed must respect the rights, traditions, traditional knowledge, and cultural practices of Indigenous peoples.

According to a 2021 Indigenous Climate Action report, *Decolonizing Climate Policy in Canada*, Prime Minister Justin Trudeau's attempts at climate change policy frameworks in 2016 and 2020 fell short of meaningfully incorporating the rights and needs of Indigenous peoples, while continuing to create inequities and inequalities among the most vulnerable.[28] The report found the needs and voices of Indigenous peoples were merely tokenized. "Indigenous Peoples and our rights, knowledge, and climate leadership were mentioned again and again in both plans, yet we were structurally excluded from the decision-making tables where these plans were made." Indigenous Climate Action found both frameworks allowed the continuation of fossil fuel production, and the solutions proposed ignored "the realities faced by Indigenous communities and Nations, and fail to address the structural inequalities continually reproduced through ongoing colonial relations and policies in Canada."[29] Though this analysis came before the 2022 release of the Liberals' 2030 Emissions Reduction Plan, a similar critique could no doubt be made of it. As Indigenous Climate Action writes, "colonialism caused climate change," and in order to tackle climate change, we need to transition away from "growth-driven capitalism" and carbon-intensive economies.[30]

It is Indigenous peoples who live closest to the land and these life-giving forces. It is our responsibility to change the trajectory of the path the world is on and begin disrupting the systemic oppression and environmental racism towards Indigenous peoples, by affirming our knowledge systems and ways of knowing, being, and doing as the thread that weaves our fabric of life. The basis of Indigenous peoples' development comes from inquiry through doing, practising, exercising, and experiencing knowledge and interconnectedness with the land. Promoting these types of inquiries will be the beginning of a system and discourse change in education, policy, and laws.

## HOPE FOR SOVEREIGNTY AND GOOD LIVELIHOODS
### Angele Alook

I begin my story in a painful place in order to come to a hopeful place for my people and the land. On May 8, 2012, I woke up crying at my father's house on Bigstone Cree Nation reserve 166D. I went outside and sat in a clearing in the bushes behind the house, where my father had arranged a circle of log stumps. Sitting on one of the stumps, I stared up at the poplar trees and listened to the birds singing, mostly chickadees doing their morning songs among a symphony of little birds chirping. I had lost my brother the day before; he went out duck hunting and never came home. He died in a truck accident along a road owned by a forestry company, along a narrow path above a creek. He had gone out hunting with a cousin, and they were rushing home because they had to work early the next day out in the oil patch. Sitting out there, I thought of my brother's final moments in the early hours of the morning, dying near a creek, on the land. He died a hunter. Listening to the birds sing that morning made me angry. I prayed, and I cried and asked my Creator why these birds should live and have life, while my brother could no longer have life. In my spirit, it seemed so unfair.

Years later, as I ask questions about a just transition, I reflect on the sociological aspects of the loss of my brother. Why are our lands in Treaty 8 criss-crossed with forestry company roads and oil company roads, and pipelines and oil leases, where you must "drive at your own risk," where you are on company property? I think of recent conversations I've had with my mom about how one of my grandfathers was hit by a logging truck, and how we've lost young people in accidents from the constant industrial vehicles pulling natural resources out of our territory. I think of the stress working in the oil patch had on my brother, and the capitalist need for profit that meant he should work long hours, and the reason they rushed home on that precarious road. I think of the interviews I've done with people in my community for research with the Just Powers project, a collaboration based at the University of Alberta focused on creating socially just approaches to energy transition and more livable futures for all. I spoke with trapline owners and traditional knowledge keepers, whose traplines were encroached on by pipe-lines, roads, powerlines, and forestry stockpiles, and the cumulative impact all this resource extraction had on their hunting and harvest-ing rights, on their medicine and berry patches, and on intergenera-tional teachings on the land.

I also think of the hope my brother has left me with, the hope for a better way, the hope for returning to our traditional ways of living and being on the land. Towards the end of his short life, my brother had begun taking up the role of a hunter, learning to hunt various types of game and fowl all over Treaty 8, from cen-tral to northern Alberta, up to the Northwest Territories and into northeastern British Columbia. He was always coming home and sharing out the meat with Mom and aunties and nohkomak (grandmothers).

At the end of each of my interviews with elders and traditional knowledge keepers for the Just Powers project, I asked: What are your hopes for the future? The answer for most was simple. Their hopes for the future were that our children and grandchildren could hunt and harvest, and live and learn on our land, as our people have done since time immemorial. The hope for clean land and water. Hope for our children and grandchildren to have sovereignty on the land, as stewards of the land. Our colonial history has brought so much death, violence, genocide, and continuous traumas, that sometimes hope is hard to come by. But when you ask Indigenous peoples about their hopes, it is a hope for life, for a good life, for good livelihoods, a hope for the respect of Indigenous rights, and hope for sovereignty.

## Land Back?

Land Back is a new slogan, originating from Indigenous artists and activists online, reiterating the long-standing demand for the abolition of colonialism. As stated in the Yellowhead Institute's *Land Back* Red Paper, "in order to more fully regain and exercise self-determination generally, Indigenous people require significant economic bases and sources of revenue to pull out of generations of systemic impoverishment. This is also a matter of economic justice."[31] As nehiyaw-iskwewak, Angele and Crystal have seen the Crown and its corporations enter our lush boreal forest territories to access our natural resources without regard for the environment or our inherent and Treaty rights. It is difficult in the short term to imagine Canada giving us our land back, since the Canadian government maintains that its sovereignty is superior and it continues to deny the true spirit and intent of Treaties. Add to this that many Nations have come to rely on the extractive industries in our backyards. Yet a just transition is not possible without radically remaking these relationships, and we see hope in the movements advocating for Indigenous and Treaty rights and the people defending our lands against further destruction.

Indigenous peoples getting land back means allowing us to return to our traditional livelihoods, governance, languages, and cultures. It means abolishing colonial and capitalist systems. As nehiyaw legal scholar Sylvia McAdam (Saysewahum) says, "'Give it back' means to restore the livelihood, demonstrate respect for what is shared—the land—by making things right through compensation, restoration of freedom, dignity, and livelihood."[32]

Most importantly, Land Back is about reclamation. As Mike Gouldhawke, a Métis and Cree writer and social movement organizer, writes, "The land has always been here and we've always been reclaiming parts of it. So, Canada's challenge is how to keep us off of it and how to keep us from holding onto the idea that it's right for us to reclaim it." "Cree laws," writes Gouldhawke, "like sihtoskâtowin (coming together in mutual support) and miyo-wîcêhtowin (the intentional cultivation of good relations) stand in stark contrast to this settler system, which is based on private and individualized rights to property and political representation." He adds:

It is essential for our relations to grow as they need to be strongly rooted in the material and spiritual reality in which we live, on our territories (cities are also part of this land), and in solidarity with other oppressed people who are also struggling, reciprocating the solidarity they have shown to us in our times of resistance.[33]

## Return to Treaty: Upholding Sovereignty

The calls from Indigenous movements for Land Back are consistent with assertions of sovereignty, which have been ongoing since contact. We have long struggled to get Canada to recognize and implement our sovereignty and our authority over our lands and our Nations. To repeat, these are our inherent rights, rights that were affirmed and added to in the process of Treaty making. Yet Canada and the provinces have failed time and again to implement the Treaties and to respect and uphold our rights. Indigenous peoples face a myriad of disadvantages due to the theft of our lands and resources and the continuing attempts by successive governments to deny us our lives, our children, our cultures, and our lifeways. Indigenous peoples need a system built for us and by us and that serves us—we don't need to be assimilated into Canadian society and laws. None of this is possible without inherent and Treaty-based frameworks at the forefront of the advancement of our sovereignty and self-determination.

From an Indigenous perspective, we see a pressing need for accessible traditional ecological knowledge, and for the presence and practices of traditional Indigenous wellness knowledge among all professionals who work in Indigenous Nations and with Indigenous peoples. First Nations must move away from colonial systems that support unequal power relationships and threaten their inherent and Treaty rights. These colonial processes continue to undermine the Treaty relationship and continuously set unrealistic timelines for our Nations and leaders, without providing adequate resources to help our leaders to make informed decisions. As First Nations we require initiatives that support Treaty-based funding processes, models, and frameworks, in order to address the effects of the persistent attempts to colonize the first peoples of this land.

We need to prepare for the end of this world, an end to the extractive capitalism and colonialism that cause climate change, returning to Treaty and recognizing and upholding Indigenous sovereignty. According to the Indigenous Environmental Network, as Indigenous peoples, "a Just Transition recognizes our Indigenous rights, sovereignty, and assertion of self-determination to control and manage our ancestral lands, waters, and territories and all natural resources inclusive of our own laws, values, customs and traditions."[34] In this way we can begin to hope. "Recognition of sovereignty," former Manitoba Grand Chief and current Chief of the Pine Creek First Nation Derek Nepinak told us, "requires one to conceptualize the reality from beyond the confines of the social and political constructs that have been developed to hold us back from who we truly are, as empowered and enabled Indigenous peoples. If we can carry a message of sovereignty, it inspires hope, and that hope can sometimes translate into actions that are meaningful to allow people to move forward in a good way."[35]

# 2.
# DELAY AND DENY

## Canada's Approach to Climate Action and Indigenous Sovereignty

"Canada is back, my good friends," Prime Minister Justin Trudeau declared in his address at the United Nations climate conference in Paris, concluding a speech advocating for an ambitious new climate target of limiting global heating to 1.5°C. His December 2015 speech marked a shift from Stephen Harper's Conservative government's loud obstruction at these conferences for the previous decade. Trudeau's Liberals had just won a majority in the federal election that fall; the Alberta NDP, also campaigning with a climate-conscious platform, had won a majority in the spring, ending forty-four years of Progressive Conservative rule in the province. Both governments had also committed to respecting Indigenous rights and building new "nation-to-nation" relationships.

For a moment it seemed like those in power were successfully being pushed by growing social movements—including Idle No More and climate activists—to lead the way on significant change. The British Columbia NDP made similar promises when they came to power in 2017. But despite this ambitious talk, not much has fundamentally changed. Indigenous rights continue to be violated and emissions keep creeping up. In this chapter, we ask how and why this has happened. And we give a longer view of how Canadian governments and industry leaders have, over the past fifty years, acted to dampen or outright block most popular proposals to build a different society. The first step of breaking out of this pattern is to learn about it.

### Social Movements Grow

The social movement forces at play today, though they are part of a trajectory that started with European contact, largely took shape in the decades of the mid-twentieth century. The Second World War (1939–45) both ended a ten-year-long economic depression and disrupted life as people knew it. The intense sacrifices made during the conflict created an opening for many people to reconsider what kind of society they wanted after the war, and to fight for it. Many workers in what's known as North America unionized and took mass strike action to win social safety nets, public housing, and safety standards. Feminist groups fought for the right to birth control, equal pay, and more; the Black Power, Red Power, and Chicano movements were battling colonialism and structural racism; and the LGBTQ2S+ movement was in ascendance.

This was also a time of rapid industrialization, along with urban and suburban expansion, which meant massive growth for chemical, coal, oil and gas, and car companies. More and more people were also noticing the impacts of these industrial activities. In the 1950s, farmworkers in California were speaking out about pesticides and their health impacts, the Dene Nation and workers at the Giant gold mine in the Northwest Territories were calling attention to arsenic pollution, and scientists were studying the effects of the vast array of relatively new industrial substances.[1] For Indigenous peoples, it was nothing new to be connected to and care for lands, waters, and non-human beings, but for the large settler population, a mass environmental consciousness was just starting to take shape. For centuries, the so-called Americas had been seen by many settler leaders and thinkers as nearly inexhaustible in their abundance, with endless frontiers to exploit.[2]

Prominent works—like Rachel Carson's 1962 book *Silent Spring*, which exposed the damages of pesticides—challenged assumptions about unending industrial growth. The new consciousness developed and spread, for example with the founding of the peace- and environment-focused activist group Greenpeace in Vancouver in 1971 and the publication of the Club of Rome's *Limits to Growth* report in 1972. Under pressure from an increasingly organized and informed public, governments in the United States and then Canada created environmental agencies and passed some of their first pieces of environmental legislation in the early 1970s.

Indigenous peoples in the 1960s were organizing and politically educating one another as part of the Red Power movement. (By some accounts, this term was coined by the National Indian Youth Council in the United States, while others attribute it to Vine Deloria Jr. of the Standing Rock Sioux.) The Red Power movement confronted colonial control and dispossession of Indigenous peoples, including environmental destruction. At this time, some of Canada's most oppressive, genocidal policies and activities had just been lifted, but many still remained. In Canada, the pass system, which kept Indigenous peoples on reserve lands unless a settler bureaucrat (an Indian Agent) allowed them out, had been in effect from the 1880s until the 1930s and 1940s. Indigenous peoples could not hire lawyers to pursue land claims from 1927 to 1951. Residential schools were in operation in the 1960s and in some places for decades to come, and the Sixties Scoop of Indigenous children being taken from their families was ongoing, often to disastrous effect.[3] The Canadian and US governments had enacted policies in the 1950s to push Indigenous peoples off reserves and into urban environments to assimilate into settler society, part of a project to weaken and ultimately end government-to-government relations with First Nations, Native tribes, and other Indigenous governments. This led partly to what the settler government called the "Indian problem," the supposed problem being Indigenous peoples migrating to urban centres, raising issues of whether they would assimilate into urban labour markets, economies, and social institutions—a social problem premised on racial stereotypes of the "drunk Indian," "lazy Indian," and "welfare Indian."

A new round of Indigenous resistance began in 1969, after Jean Chrétien, the Liberal minister of Indian Affairs (as the department was then called), and Prime Minister Pierre Elliott Trudeau proposed ending Indian status and thereby ceasing to recognize Indigenous peoples. This proposal, termed the White Paper, would have refused to recognize inherent rights and eliminated the Canadian state's agreements with and obligations to Indigenous peoples, like those agreed to in the Treaties (while at the same time dealing with the "Indian problem" of assimilation). A crucial implication was that the Canadian government would have been able to consider all land within so-called Canada unquestionably its own, as all rights held by Indigenous peoples would be extinguished.

Put forward after months of consultation with Indigenous peoples, the White Paper marked a major betrayal of what had been discussed. Many Indigenous peoples were furious. As Cree Chief Harold Cardinal puts it in his famous 1969 response, a book titled *The Unjust Society*:

> Small wonder that in 1969, in the one hundred and second year of
> Canadian confederation, the native people of Canada look back on
> generations of accumulated frustration under conditions which can
> only be described as colonial, brutal and tyrannical, and look to the
> future with the gravest of doubts.[4]

He speaks also to the many Indigenous peoples finding their voice in
that historical moment, writing, "Faced with society's general indiffer-
ence and a massive accumulation of misdirected, often insincere efforts,
the greatest mistake the Indian has made has been to remain so long
silent."[5] Cardinal and many other Indigenous activists fought back,
and their opposition to the Liberals' proposal became impossible to
ignore. After a massive mobilization effort, the White Paper proposal
was dropped in 1971.

In addition to Indigenous mobilizations on the ground, a number
of precedent-setting legal decisions began to undermine Canada's claims
to sovereignty. The Calder case, brought forward by Nisga'a Chief Frank
Calder on the west coast in 1967, resulted in a Supreme Court ruling in
1973 that called into question Canada's claim to vast swaths of land and
pointed to the legitimacy of Indigenous title to these lands.[6] Around this
time, several Indigenous leaders were also building connections and alliances with leaders involved in self-determination movements in Africa
and Asia, strengthening the global anti-colonial movement.

The beginning of the 1970s marked the end of the postwar economic
boom. In search of new sources of profit, settler capital aggressively
pushed into Canada's North in search, particularly, of fossil fuels, hydro-
electricity sources, and minerals. This led to head-on collisions with
Indigenous communities just as they were asserting their sovereignty in
a new round of mobilizations. In the Arctic in the 1970s and 1980s, Inuit
struggled against the Polar Gas Project and Arctic Pilot Project, which
aimed to extract and transport huge quantities of fossil gas from the
region. The Dene Nation's resistance to the Mackenzie Valley Pipeline
proposal in the 1970s in the Northwest Territories was a major struggle
over control of land and resources. Indigenous governance scholar Cliff
Atleo points to testimony given at the Mackenzie Valley Pipeline Inquiry
(or Berger Inquiry, as it was led by Justice Thomas Berger) as giving
inspiration to generations of Indigenous activists.[7] In particular, he cites

Frank T'Seleie speaking in 1975 as Chief of the Fort Good Hope Dene Band, located along the Mackenzie River in the Northwest Territories.

> Whether or not your businessmen or your Government believes that a pipeline must go through our great valley, let me tell you, Mr. Berger, and let me tell your nation, that this is Dene land and we the Dene people intend to decide what happens on our land. . . . There will be no pipeline because we have our plans for our land. There will be no pipeline because we no longer intend to allow our land and our future to be taken away from us and that we are destroyed to make someone else rich. There will be no pipeline because we, the Dene people, are awakening to see the truth of the system of genocide that has been imposed on us and we will not go back to sleep.[8]

While Indigenous peoples were saying no to these colonial extractive projects, as Dene scholar Glen Coulthard explains, these statements "also have ingrained within them a resounding 'yes': they are the affirmative *enactment* of another modality of being, a different way of relating to and with the world."[9]

Many settlers remain in denial that the land in Canada is unceded, but the truth, as we saw in the previous chapter, is that the Treaties were agreements to share the land, not surrender it. And much of the land, like the majority of British Columbia and vast swaths of Canada's North, had no Treaties describing how it would be shared. This period around the 1970s also brought an awakening for some settlers and the beginnings of widespread solidarity between movements. Vern Harper, a leader of the Canadian section of the American Indian Movement spoke about organizing in this time:

> One of the key factors that made [this] a turning point was that native activists, for the first time in their generation, realized that there was non-native support for their cause. The isolation of the natives, used by the state, is no longer effective. . . . We see trade unions, progressive left groups, church groups such as the Quakers, even liberal elements give support, such as funds, telegrams, participating in demonstrations, letters to Members of Parliament denouncing the tactics of the state, to help us.[10]

## Industry Responds:
## Canada's Backroom Approach to Indigenous Rights

It has always been important for the Canadian government and its supporters in business to have certainty about who controls the land. As
far as the leaders of resource industries were concerned, the growing
power of the environmental and Indigenous rights movements represented threats to their business. And Indigenous rights assertion was
of special concern to the Canadian government because it undermined
the legitimacy of Canada's claim to jurisdictional supremacy. As these
movements grew, industry and its close allies in government looked for
ways to neutralize the movements' militancy, so that favourable conditions could be maintained for settler businesses and private wealth could
continue to accumulate.

Canadian governments and industry didn't abandon their efforts
after Indigenous peoples defeated the White Paper, won the Calder case,
and made other gains. Kahnawake Mohawk policy analyst Russ Diabo
points to internal government correspondence at the time showing that
the federal government planned "not to abandon the White Paper Plan,
but to change tactics."[11] As Diabo writes:

> From the Indian Act to the 1969 White Paper on Indian Policy, to
> the "Buffalo Jump of the 1980's" [a leaked government document
> again describing a plan to extinguish Indigenous rights] to the suite
> of legislation from the Liberal government of Jean Chretien and
> now the [Justin] Trudeau government's 10 Principles, dissolving the
> Department of Indian Affairs and the "Framework" to define "rights"
> in legislation, the government of Canada has been implementing a
> plan to terminate our collective Aboriginal and sacred, historic Treaty
> rights.[12]

These new strategies continue the aim of securing control over land and
resources, but they have been cloaked in increasing lip service to partnership, consultation, and other strategic words.

Drawing on the work of Diabo and others, we can identify three
keys to the federal government's long-term strategy to maintain full control over the land: the establishment of a modern treaty process, "self-
government" agreements, and a focus on "consultation" around industrial development that increasingly results in impact benefit agreements

and business investments and partnerships. These strategies persist into the 2020s.

Briefly, modern treaties are agreements where Indigenous peoples surrender large parts of their traditional lands to the Canadian state in exchange for some land, rights, and other benefits, usually including cash. Modern treaties, also known as comprehensive land claims, are pursued where historic treaties do not exist. Though these agreements generally mark an improvement from the state-administered poverty of the reserve system, they are goal-defined by the Canadian state and require a surrendering of rights to large swaths of traditional territory. The first modern treaty came into effect in 1975, the now-famous James Bay and Northern Quebec Agreement signed by James Bay Cree and Northern Quebec Inuit representatives.[13] Negotiations of modern treaties are very slow and costly, and because they present the possibility of a new "nation-to-nation" relationship, they have tended to dampen or prevent more militant action like protests and blockades in the territories under discussion. Several modern treaties have been signed, covering huge areas, especially in so-called Canada's North. The creation of the newest territory, Nunavut, was the result of a modern claims process.* The vast majority of British Columbia, on the other hand, has no Treaty agreements, despite decades of persistent and often coercive efforts from successive federal and provincial governments to negotiate them.

On lands where Indigenous peoples and settlers entered into Treaty before 1923, Canada uses a written record to claim it has ultimate control over the land. Even on reserve lands, Canada claims jurisdiction and only recognizes limited rights on Nations' much larger traditional territories. Those rights usually include Indigenous peoples' right to hunt, trap, fish, and gather. However, individual settlers and colonial police have consistently restricted even those limited rights, demanding to see provincial

---

* In 2022, Kunuk (Sandra) Inutiq, former director of self-government at Nunavut Tunngavik Inc., the legal representative of the Inuit of Nunavut for the purposes of Treaty rights and Treaty negotiation, wrote of governance within the modern treaty framework, "The assumption that the public government is a form of self-determination is a farce: Inuit interests have not been served by the government because Inuit lives have not improved. What we have is a façade of self-determination where senior bureaucracy is made up of 85% non-Inuit upholding a system that benefits settlers." Kunuk Inutiq, "Hungry Days in Nunavut: The Façade of Inuit Self-Determination," Yellowhead Institute, May 27, 2022, yellowheadinstitute.org.

fishing licences from Indigenous fishermen, strengthening trespass laws so that Indigenous peoples have to ask permission to access "private" property, and criminalizing and using violence against those who exercise their rights. On a larger scale, governments have felt emboldened to infringe on rights whenever they deem it in their interest, for example, to set up dams or open-pit bitumen mines, destroying traplines and ecosystems in the process. Indigenous Nations' efforts to assert rights in this context have sometimes involved land defence actions, like the logging road blockades Asubpeeschoseewagong (Grassy Narrows) First Nation, located in Treaty 3, northwestern Ontario, set up to protect their forests. The courts are another option; the Beaver Lake Cree Nation, for example, has for well over a decade been pursuing its claim for Treaty violations due to the cumulative impacts of industry on its traditional territory.

Colonial governments have tried to diffuse these assertions of rights among historic treaty signatories by offering to recognize general, unspecific rights and enter into "self-government" agreements.[14] As Diabo describes it, self-government agreements sound good, since they give Indigenous Nations the ability to govern certain aspects of their society, such as child and family services. But cloaked in these agreements is often legal language that effectively turns signatories into municipalities (or sub-municipalities) and binds them not to exercise rights to their traditional territories. In conjunction, Canada claims to repatriate lands to Treaty Nations by turning them into "fee simple," meaning private property, first under control of Band Councils. These lands become subjected to colonial federal, provincial, and municipal laws, rather than being communally held sovereign lands.

Consultation, the third prong of the colonial approach, has been pursued by governments and corporations that resist acknowledging the full reality of Indigenous rights but are keen to avoid legal battles, protests, and blockades. Industry and government's consultation processes, on oil sands mines or pipelines, for example, are often only as rigorous as First Nations force them to be. When a First Nation resists, the favoured approach of fossil fuel companies has been to engage in isolated, quiet negotiations with Band Councils. These Band Councils are legal creations of the federal government's Indian Act, and in many cases displaced traditional, existing Indigenous governments. Band Councils are dependent on the federal government for their budgets, budgets that are often so meagre that Band Councils find themselves in the difficult position of being administrators of poverty in their own communities. This makes offers from industry appealing. And these offers typically take the form of confidential benefit agreements where, in exchange for a relatively small amount of money and temporary jobs, the federally imposed Band Council agrees to support the project. The agreements typically stipulate that Band Councils will try to prevent community members from opposing the industrial development.

When First Nations oppose resource projects like pipelines—whether through the Band Council or traditional systems of government—colonial courts often rule against them, saying that contrary to what they are experiencing, their rights, as understood by Canada, are not being infringed upon. This was the case in 2020 when the Tsleil-Waututh Nation, Squamish Nation, and Coldwater Indian Band opposed the Trans Mountain Expansion Project oil sands pipeline; as it was in 2017 when the Chippewa of the Thames First Nation opposed the reversal of Enbridge's Line 9 pipeline in Ontario. The government of Canada and pipeline companies did not respect these Nations' right to free, prior, and informed consent, as laid out in the United Nations Declaration on the Rights of Indigenous Peoples (UNDRIP), which Canada had adopted in 2016. But when challenged, the Supreme Court of Canada did not overturn these infringements of rights, allowing the projects to continue.[15]

Court cases and modern treaty negotiations take years, even decades, and sometimes do not deliver clarity one way or another. Negotiations can break down, and colonial courts sometimes affirm Indigenous rights exist, but direct the parties to negotiate what those rights are in practice. That was the outcome of the landmark 1997 Delgamuukw Supreme Court case, brought forward by Gitxsan and Wet'suwet'en peoples, where it was affirmed that they had never ceded a massive 58,000-square-kilometre territory (in northern British Columbia), but the judges then instructed the parties to negotiate a resolution. That resolution has still not happened.[16] While these cases and negotiations go on, industry continues extracting resources from Indigenous lands.

Injunctions are a short-term measure often used by industry when Indigenous peoples resist their activities. Secwepemc leader Arthur Manuel referred to injunctions as "a legal billy club" to forcibly remove Indigenous peoples from their own land, often using the RCMP.[17]

Because settler governments assume the land to be theirs until proven otherwise, Canadian courts deem activities that impede business to be illegal. The Wet'suwet'en, for example, have blocked roadways used for the construction of the Coastal GasLink pipeline, in their territory where the Supreme Court has affirmed they have rights to that land. In response, hundreds of RCMP officers were deployed, including those conducting "lethal overwatch," meaning the deployment of an officer who is prepared to use lethal force, to enforce the pipeline company's injunction in Wet'suwet'en Yintah (territory) in 2019.[18] Raids were repeated in 2020, and 2021, and the RCMP maintained a significant presence there in 2022, conducting near-daily patrols of camps.

Through modern treaties and various methods of rights denial, settler governments push legitimate disagreements into backrooms and courtrooms, with the implication that any other forms of asserting rights on the land are inappropriate. Far from respecting the need for consent, settler governments have used a seemingly endless array of tools to avoid the possibility that Indigenous peoples could stop, or veto, industrial activities. But Indigenous resistance continues undeterred on the land. Settler politicians keep pointing to the courts, the colonial "rule of law," and negotiating tables while claiming they respect Indigenous rights and are acting in line with vague concepts like "reconciliation." And this approach of saying the right thing in public while doing the opposite in the halls of power is not unique to Indigenous rights. It closely echoes the way Canadian governments have approached environmental and climate regulations.

## Sustainable Development Discourse and Voluntary Climate Measures

The first United Nations Conference on the Human Environment (the Stockholm Conference) was held in 1972.[19] While it's not obvious from the name, industrial business interests had a robust presence at these conferences from the beginning. Chairing the first conference was Maurice Strong, a former Alberta oil executive with deep connections to the Liberal party and the Canadian business community. Industry, seeing the calls that were arising at the time for tough environmental protections as a potential impediment to profits, was looking for ways to manoeuvre. In Stockholm, Strong advanced the idea that "there is no fundamental conflict between development and the environment." This is strikingly similar to Justin Trudeau's oft-repeated maxim fifty years later that "the economy and the environment must go hand in hand." The statements are not philosophically untrue; there are ways to design an economy that meets people's needs while stewarding land and life. But this framing has consistently been used instead to legitimize a very specific form of economic development with little regard for environmental concerns or human well-being, as has been described in detail in the books *The Big Stall* by Donald Gutstein and *The Trudeau Formula* by Martin Lukacs.

The ideology Strong and others were pushing for years at the domestic and international level largely crystallized around the term "sustainable development." The term really caught on following publication of *Our Common Future*, the 1987 report of the Brundtland Commission, a UN organization continuing the international-level work launched at the Stockholm Conference. The seductive idea behind sustainable development—at least the version put forward by politicians and industry globally—is that the current unfair distribution of resources and levels of pollution within and between countries are not problems to be addressed directly by limiting industry's growth and redistributing wealth so that everyone can live well and meet their needs. Instead, the proposed solution is for continuous economic growth—to keep producing and consuming more and more materials and energy, but with promises of greater efficiency and corporate responsibility at some later date. These measures, supposedly, will rein in negative impacts and ensure profits and benefits trickle down to the global poor. Plastics, for example, can continue to be produced in ever-greater quantities, with promises of better recycling systems and community development, someday, even as landscapes (in poorer areas) and oceans fill with plastic and economic inequality only grows. The effect in the near term is that the economy as it currently exists, along with the associated negative impacts, is not fundamentally questioned or altered.

The stated rationale for this conception of sustainable development was that economic development was needed in the Global South (or "Third World"), and so pollution was justified, as it was part of raising standards of living. In practice it was adopted and mainstreamed by Global North countries like Canada looking for new outlets for foreign investment, and thus profit making, abroad. At home, sustainable development was the corporate challenge to the idea of ecological limits, which had been so fundamental to the environmental movements of the 1970s. Instead of fashioning economies within the earth's limit, sustainable development promised indefinite economic growth as long as it continued to improve efficiency. Multinational corporations based in Global North countries have used this philosophy to help justify resource extraction in countries of the Global South, even though the profits from these activities overwhelmingly flow out of the country into corporate coffers. For example, 75 percent of global mining companies are headquartered in Canada, and many have a documented history of committing horrendous human rights violations and polluting local environments around the world, while at the same time touting their corporate responsibility.[20]

This type of "sustainable development" suits resource industries just fine, as long as there are no strict limits and regulations. To that end, the oil companies and their friends in politics have worked to ensure that emission reductions are a voluntary part of international agreements, and that targets are non-binding and without penalties for non-compliance. Industry got its way in the Kyoto Protocol on climate, signed by Liberal prime minister Jean Chrétien in 1997 and ratified by his government in 2002, and in the Paris climate agreements, which Liberal prime minister Justin Trudeau committed to in 2015 and signed in 2016. The targets were voluntary and non-binding, and there were no penalties if industries or governments missed them. This has allowed industry and Western politicians to circumvent the UN Framework Convention on Climate Change's founding principle of "common but differentiated responsibility," which says targets for wealthy countries should be stricter according to their contribution to the crisis and ability to pay. Voluntary and non-binding targets make it easy to stay unaccountable.

In Paris, there were in effect two agreements made. One was the big public agreement to limit warming to 1.5°C, and the other was that countries would set voluntary targets that, taken together, would not come close to limiting emissions to 1.5°C. There was also a third unspoken agreement: not even the inadequate voluntary targets would be enforced, so industry would hardly be affected by the agreement and could continue with business as usual, as long as it was dressed in far-off commitments to reduce emissions. Politicians in the Global South, Indigenous leaders, and climate activists protested these voluntary targets, instead advocating for mandatory limits. But these dissenting groups were locked out of negotiations for the most part, while the fossil fuel industry was invited to draft the agreements.

At the COP26 climate conference in Glasgow, Scotland, in 2021, fossil fuel companies were, for the first time, not formally invited inside to participate. However, reports suggest the industry was still able to exert influence on proceedings through friendly politicians, and consultants hired to do their bidding. These consultants were ultimately a larger group than any single country's delegation, outnumbering the event's official Indigenous constituency by around two to one.[21] "Fossil fuels" made it into the official agreement text for the first time (having avoided mention in previous agreements), but no significant movement was made towards putting adequate and mandatory emissions reductions in place. The ambitious-sounding targets announced were, as usual, voluntary, not differentiated to ensure the wealthiest countries and companies do and pay their fair share, and without tough enforcement mechanisms.

In addition to blunting environmentalists' demands, the corporate framing of sustainable development also fulfills another goal of industry: bringing workers onto their side. Workers in oil, gas, and especially coal have, throughout the decades, been in conflict with their employers over working conditions and pay. Why, workers asked, were they put at such risk and paid relatively little, compared to the companies' huge profits? Workers' grievances, paired with concerns that the industries were polluting the earth, could result in militant labour groups and other social movements, like environmental movements, working together to oppose the industry's plans. As early as the 1940s, forestry workers in British Columbia who were doing dangerous jobs were raising the issue of deforestation as both a long-term threat to those jobs and a threat to the natural environment.[22] That sort of opposition, if well organized, could get in the way of growth and profits. To avoid this outcome, the fossil fuel industry has created advocacy campaigns targeted at developing fossil fuel workers as defenders of their industry.

Social media campaigns organized by Canada's Energy Citizens (a campaign of the Canadian Association of Petroleum Producers), Oil Sands Action (a campaign with more grassroots beginnings but that has since been accepted oil and gas funding),[23] and others have repeated the message that the resource industries are working hard to clean themselves up and be socially responsible, that they are providing necessary sources of materials and energy and relatively well paying employment, and that environmentalists and Indigenous peoples asserting their rights are a threat to workers' livelihoods, because these activities threaten the industry.[24] Industry's opponents are presented as powerful, unreasonable radicals. These lines of thinking are pushed by industry-funded media, social media campaigns, conservative parties, and strategic philanthropy by resource extraction companies.[25] The effect has been that in Alberta, Saskatchewan, and other oil and gas producing regions in Canada, workers in that industry are some of its fiercest defenders. However, this outlook is certainly not uniform and there are dissenting views within the workforce.

### Neoliberalism and Colonial Extraction

As sustainable development discourse was forming, the oil and gas industry was growing, and who controlled it was changing. In the early 1970s, American-owned companies dominated, but the federal and provincial governments in so-called Canada were embracing economic nationalism, part of an international trend, and saw opportunity in the large reserves of oil and gas here. To spur on the industry, and to increase Canadian and public control of it, governments set up oil companies in the Arctic (Panarctic, 1966), Saskatchewan (SaskOil, 1973), Alberta (Alberta Energy Company, 1973), and Canada (Petro-Canada, based in Calgary, 1975).* These Crown corporations and joint ventures developed new technologies for producing unconventional (heavy and bituminous) oil and opened up new extraction projects. They also gave governments a window to see directly how industry works. And notably, they delivered profits to governments.

Seeing these Crown corporations as a threat to private investment, business-backed forces in Canada sought to mount a counteroffensive. This was in the context of the Cold War; communist and socialist movements had been making gains around the world, and right-wing governments and business interests were seeking to limit and reverse that tide. In many Global South nations, the response involved US-backed military coups and bloody dictatorships that implemented business-

---

* The first president of Petro-Canada, in 1975, was Maurice Strong, who had chaired the first United Nations Conference on the Human Environment in 1972 and stayed involved with the conferences for decades.

friendly economic reforms, for example, in Indonesia in 1965, Chile in 1973, and Argentina in 1976. The economic system they imposed is often described as neoliberalism, a form of capitalism characterized by privatizing formerly public services and utilities, removing regulations and cutting taxes for businesses, and cutting remaining public services as a way to force people to accept low-wage work with poor conditions to survive. Neoliberal ideology also tells us that everyone is an individual entrepreneur, only looking out for themselves. As UK prime minister Margaret Thatcher put it, "There is no such thing as society"—only individual responsibility.

Neoliberalism came home to Canada and the Global North later and with less overt violence than in Global South nations. Here, governments enacted legislation that significantly altered society, while supportive public intellectuals and media figures provided ideological justification for the changes. Canada's neoliberal turn began in the mid-1970s during Pierre Trudeau's time as prime minister, but it really took off under Progressive Conservative prime minister Brian Mulroney, who was in power from 1984 to 1993. Under Mulroney, then Liberals Jean Chrétien and Paul Martin, wages and workplace safety came under attack, as did funding for healthcare, education, unemployment support, and other programs won by social movements in the 1960s and 1970s.

For oil and gas, neoliberalism took the form of privatizations and deregulation. Mulroney initiated the privatization of Petro-Canada, and Chrétien's Liberals and successive governments continued the process. Petro-Canada was eventually fully sold off and as of 2022 was part of Suncor, one of the largest, most profitable Canadian oil and gas companies. The provincial conservative governments of Grant Devine in Saskatchewan (1982–1991) and Ralph Klein in Alberta (1992–2006) likewise privatized public oil and gas assets. After governments had invested to expand the industry, the private sector reaped the rewards and looked to continue expanding. Industry also got the regulations it wanted to do just that. As Donald Gutstein writes in *The Big Stall*:

> March 6, 1996 was "a beautiful day for bitumen," industry magazine *Oilweek* effused. Not only was there no carbon tax in [Liberal] Finance Minister Paul Martin's budget, tabled that day in the House of Commons, but the industry received what it wanted—a universal low-tax regime for all oil sands producers.[26]

This suite of privatizations and deregulation represented a substantial devolution of control of the industry from the public sector to private investors.

The neoliberal turn, beyond enforcing suffering to bolster corporate profits, was framed as the new common sense. This has deeply constrained the political imagination in Canada. Our ability to think big about climate action, Indigenous sovereignty, and wealth redistribution has become severely limited. Since the 1980s and 1990s, most major federal and provincial parties, including the NDP, have openly competed in elections to show who can best balance budgets, demonstrate fiscal stewardship, and protect private profits. Still, inspiring opposition to this consensus has persisted, a thorn in the side of the colonial, neoliberal offensive.

### The New Climate Denialism, or Climate Delayism

As the world has learned about the dangers of climate change publicly, beginning in the 1980s, and as many people in so-called Canada learn about Indigenous rights, the oil and gas industry has adapted its tactics to continue pursuing its business interests. Thankfully, the time has largely passed when outright denial of climate change could get airtime in mainstream media. That era lasted far too long, extending well into the 2000s. It was crafted and perpetuated strategically by various companies, most notoriously by oil giant Exxon, in collaboration with industry front groups and think tanks, as the #ExxonKnew campaign has documented. Exxon's Canadian subsidiary Imperial Oil employed the same tactics here.[27]

But the new climate denialism, or climate delayism, is no less dangerous. As outlined by Seth Klein and Shannon Daub at the Canadian Centre for Policy Alternatives, in this approach, the fossil fuel industry and political leaders assure us that they understand and accept the scientific warnings about climate change, but they continue to deny what this scientific reality means for policy and they block necessary large-scale actions to address the problem.[28] This approach has taken different forms over the years; in Canada, it really arrived with carbon pricing proposals.

In 2007 climate was in the spotlight, and social movements were mobilizing. Al Gore's massively popular climate documentary *An Inconvenient Truth* had come out the year before, and the Intergovernmental Panel on Climate Change (IPCC) had just released a report forecasting climate chaos if business as usual continued. That same year, the Business Council of Canada proposed putting a price on carbon. The Business Council is made up of the chief executive officers of the biggest businesses in Canada. And while the Business Council usually flies under the radar in Canadian media, it exerts enormous influence on the direction of economic policy in Canada and therefore on the direction of the country as a whole. Why was this high-powered group, including oil and gas executives, proposing a price on carbon, often known as a carbon tax? And why is it a form of climate denialism?

The Business Council was keenly aware that public opinion was shifting and a credible-seeming climate framework was needed in Canada to give legitimacy to ongoing business activities. A price on carbon would not put a hard cap on emissions, a major advantage for businesses. The Business Council also wanted to influence policy in favour of the oil and gas industry, writing in 2007 that "[energy-intensive industries] need a stable and predictable policy regime that provides long-term guidance and does not disadvantage investments already made."[29] They saw a price on carbon as key to allowing the fossil fuel industry to proceed with increasing production and profits.

But is carbon pricing effective? In 2007 it was relatively untested, its legitimacy based on the modelling work of economists. What was clear then was that relatively low prices on carbon would have relatively minor effects on emissions, meaning that in the short term, little would change. The promise was that the price on carbon would escalate higher and higher, the policy British Columbia and then Canada eventually adopted, with the idea that this would have a greater and greater effect. But in many cases that momentum stalled. When it introduced the measure in 2008, British Columbia was touted as a shining example of the success of carbon pricing, but the carbon price in the province did not escalate for years. And the supposed early success there, a temporary drop in emissions, may have had more to do with the coal market than carbon pricing.

Recent studies have looked at the actual effects of carbon pricing after the fact, rather than at theoretical economic modelling of what might happen. These studies have observed relatively minor effects. A 2021 report compiling thirty-seven studies found "the majority of studies suggest that the aggregate reductions from carbon pricing on emissions are limited—generally between 0% and 2% per year."[30] In some places, including British Columbia, the emission reductions attributable to carbon taxes have been cancelled out by other activities, like increases in vehicle emissions, with the overall result being that emissions have increased in the province even with carbon taxes in place. What we've seen in most jurisdictions with carbon pricing is that it has a low price bringing relatively small effects (and prompting backlash from some), yet is touted as substantial climate action.[31] All the while, the oil and gas industry increases production and emissions go up.

When the Business Council proposed carbon pricing, Stephen Harper was prime minister. Despite some indications at the time that his party planned to introduce carbon pricing, they ultimately didn't play along, possibly because Harper was ideologically opposed to a new tax, even if it was revenue neutral, meaning the government wouldn't keep any of the revenue it generated through the tax, instead rebating it all to consumers. Liberal leader Stéphane Dion campaigned for a carbon tax in his unsuccessful bid against Harper in 2008. And Justin Trudeau, who was elected as a member of Parliament that year and then Liberal leader in 2013, got the idea. In October 2013, a few months after becoming party leader, Trudeau gave a speech at the Calgary Petroleum Club, telling the high-powered crowd, "If we had stronger environmental policy in this country: stronger oversight, tougher penalties, and yes, some sort of means to price carbon pollution, then I believe the Keystone XL pipeline would have been approved already."[32] That pipeline, which was proposed to export heavy oil from Alberta to refineries as far south as Texas, was being held up at the time by US president Barack Obama's Democratic party administration (which, at the same time, was approving other pipelines and overseeing a massive boom in fossil gas and oil hydraulic fracturing, or fracking).

Rachel Notley's Alberta NDP were elected in May 2015, and Justin Trudeau's federal Liberals were elected that October. Three days after the federal election, high-ranking insiders from both governments met at a downtown Ottawa restaurant. According to a *Vancouver Sun* report, there they agreed that, in exchange for Alberta agreeing to put a price on carbon as part of a federal plan, the federal government would approve the Trans Mountain Expansion pipeline, then proposed by Kinder Morgan.[33] As the name suggests, this pipeline would help expand oil sands operations, giving access to international markets via Canada's west coast. It did not have the consent of several First Nations along the route. Trudeau then moved swiftly to lobby the US government to approve the Keystone XL pipeline. And he continued claiming he was acting on climate, making ambitious statements at the Paris climate meetings in December 2015. He also claimed, "There is no relationship more important to me—and to Canada—than the one with First Nations, the Métis Nation, and Inuit," and that "governments grant permits, communities grant permission."[34]

Around the same time, to appease oil companies in Alberta that were highly suspicious of and even actively undermining the new NDP government, the Alberta NDP put a cap on oil sands extraction. However, this cap was not going to reduce emissions. Instead it would allow for a 40 percent *increase* in oil sands production. Industry was still allowed to expand rapidly, imperilling Canada's modest Paris commitments, a fact thoroughly laid out by David Hughes for the Corporate Mapping Project. This industry expansion also implies drastic reductions in Canada's non–oil and gas sectors to make room for rapidly rising emissions from oil and gas.[35] But crucially for the federal Liberals and Alberta NDP, they could point to the supposedly reasonable new measures: the carbon tax and a production cap.

From 2015, when the Alberta NDP and federal Liberals were first elected, to 2019, bitumen production from the oil sands grew from 2.5 million to 3.1 million barrels per day, before falling slightly during the pandemic in 2020, then bouncing back up.[36] This growth and its associated emissions run directly counter to what we all know we need to do: rapidly reduce the production and consumption of fossil fuels. Claiming this policy is reasonable and adequate, as the provincial and federal governments did, is incredibly dangerous for the climate. It is new denialism.

New climate denialism has also been paired with new yet familiar approaches to denying the rights of Indigenous peoples. The Alberta NDP campaigned on implementing the United Nations Declaration on the Rights of Indigenous Peoples, then once in power, didn't. In a September 2015 interview with the Aboriginal Peoples Television Network (APTN), Alberta's new NDP premier Rachel Notley stated, "Quite honestly, we have a province that is very much driven by development and the production of our natural resources. So we're not looking at approaching this in a way that would result in economic development suddenly grinding to a halt subject to free, prior and informed consent."[37] As well, the provincial government did not pass UNDRIP legislation to make it part of provincial law. This was a capitulation to industry and (though UNDRIP legislation on its own is not a sufficient safeguard for Indigenous rights) an abandonment of responsibilities to Indigenous peoples.

As opposition to pipelines on the west coast increased, federal Liberal natural resources minister Jim Carr said in late 2016, "If people choose for their own reasons not to be peaceful, then the government of Canada, through its defence forces, through its police forces, will ensure that people will be kept safe."[38] Under pressure, Carr apologized for his heavy-handed language, loaded with Orwellian doublespeak. Justin Trudeau tried to be less forceful in his words, but still fought First Nations again and again in court to build the Trans Mountain Expansion pipeline, and watched on as Canada's national police, the RCMP, arrested those who stood in the way. To justify state violence, Trudeau invoked "the national interest," an economic argument used in that case to trample Indigenous rights and grow fossil fuel extraction. Also a case of new climate denialism.

The approach of the British Columbia NDP government of John Horgan, elected in 2017, appeared different from its Alberta counterpart, but was functionally much the same. The BC government passed an act supportive of UNDRIP in November 2019. The NDP was also support- ive, in principle at least, of First Nations opposing the Trans Mountain Expansion pipeline. However, simply passing the UNDRIP legislation did not change much in the province. The act that was passed did not magically undo all the colonial laws already in place. And it did not stop the province from pushing ahead with resource projects without the free, prior, and informed consent of First Nations, a condition within the UN Declaration. The province did not back off from the massive Site C Dam project of the Peace River, and First Nations routinely object to logging operations they have not consented to.[39]

Perhaps most stark has been the Coastal GasLink fossil gas pipeline, opposed by the Wet'suwet'en traditional government. In 2019, before the UNDRIP Act was passed in British Columbia, the RCMP raided Wet'suwet'en territory at gunpoint; after the act was passed in 2020, the RCMP were back for a similar raid in Wet'suwet'en territory. In 2021, RCMP officers aimed guns at Indigenous land defenders yet again during raids before making arrests. In early 2022, the RCMP were doing patrols through Wet'suwet'en camps multiple times a day, at all hours, without consent. At the provincial level, the Minister of Public Safety and Solicitor General, Mike Farnworth, explicitly authorized the "internal redeployment of resources within the Provincial Police Service" in a January 2020 letter to the RCMP.[40] He approved a similar deployment again in November 2021, while the province was facing catastrophic flooding in the south. UNDRIP Act or no UNDRIP Act, the British Columbia NDP were acting in the interests of the gas industry.

At the federal level, the government has also committed to implementing UNDRIP. On June 21, 2021, National Indigenous Peoples Day, Bill C-15 received Royal Assent and came into force. It took much work to secure this legislation, and a number of people see it as a victory. But Indigenous policy analyst Russ Diabo warns the legislation continues to impose Crown sovereignty, disregards Indigenous laws and legal traditions, and carries on many of the supremacist practices that have characterized the Canadian government's relationship with Indigenous peoples and their land rights in the last decades.[41]

With new denialism, which aligns with corporate sustainable development rhetoric, the oil and gas companies are increasingly touted as leaders on climate and respectful of Indigenous rights, rather than the main impediment to effective action. The packaging of new denialism keeps changing—a constant stream of new phrases and policies make it sound like the industry is changing with the times. Because most people don't have time to study the details, many are led to believe this noise indicates that significant action is being taken. But the core of new denialism persists: inadequate measures that we are told are good enough. The public gets the message that we can rest easy, knowing that leadership on Indigenous rights and climate is happening.

### "Net Zero" and New Denialism in the 2020s and Beyond

New climate denialism's face in the early 2020s has been the promise of "net zero" or "carbon neutrality." These concepts have been around for a while, but have only recently become central to the emission promises of governments and oil companies alike. Net zero, in theory, means that an activity does not overall add greenhouse gases to the atmosphere, either because the activity does not release emissions, or because the emissions made are offset by an activity that takes emissions out of the atmosphere or prevents other emissions. Take a car ride, for example. A regular combustion-engine car releases emissions, both directly from the burning of gasoline, and also because in the making of the car significant emissions were created. For a car trip to be net zero, there would either need to be some activity taken to remove carbon dioxide from the atmosphere, like planting trees, or an activity would be undertaken to prevent emissions that otherwise would have been released. Preventing other emissions could mean stopping the release of methane, a potent greenhouse gas, into the atmosphere from a source such as a landfill. It's a compelling idea, but most of these new plans for reaching net zero emissions (usually by 2050) come with dubious accounting that facilitates further delays on taking action and opens opportunities for investors to profit from the climate crisis. As Marc Lee of the Canadian Centre for Policy Alternatives details, Canada's net zero by 2050 plan is no exception.[42]

Net zero tends to rely on emissions trading, carbon dioxide removal (CDR), and carbon capture, utilization, and storage (CCUS) schemes, which often have overstated effectiveness. With emissions trading, it can be hard to know whether companies are claiming credits for doing things that they might have done anyway. Would a landfill capture its methane to use as a fuel source without being paid through an emissions trading mechanism? Would a building switch from using heating oil to geothermal and electric heating if the owner wasn't paid? These supposedly carbon-reducing actions, which governments could simply mandate, do not necessarily add up to "extra" emissions reductions. They just mean someone is paying for things that should happen anyway, and in return they get to claim their emission-emitting activity is actually "net zero."

Net zero also heavily relies on CDR technologies that are unproven and risky at a large scale. One major category of these is "nature-based solutions." Clearly, nature will be important to sequestering carbon— clear-cutting old-growth forests is a disaster for the climate, for example. And there are Indigenous-led efforts to establish conservation areas, with notable agreements made with colonial governments to protect certain areas, particularly in the North. But industry and Western nations have built an over-reliance on unproven methods of storing carbon in nature. This can take the form of planting fast-growing mono-crops of trees that displace Indigenous livelihoods. In Canada, one nature-based solution has been the Trudeau Liberals' plan, announced in 2019, to plant two billion trees in the country by 2030 as a way to sequester carbon. As of July 2022, less than 1.5 percent of that number of trees had been planted.[43] This sort of scheme, plus habitat protection, is being relied on in government modelling to sequester a huge amount of carbon in coming years. But because of increasing wildfires, pests, and other trends, Canada's vast landscapes have recently been emitting more carbon than they store, raising serious doubts about relying so heavily on these solutions.

Closely related to carbon dioxide removal is another set of technologies called carbon capture, utilization, and storage, where carbon dioxide is redirected from point sources rather than the atmosphere. This approach usually involves taking the exhaust from burning fossil fuels at industrial sites and pumping it underground. These projects have not lived up to their promised capture rates. For example, the Boundary Dam coal power plant in Saskatchewan captured only around 40 percent of the $CO_2$ it emitted in 2021 and experienced frequent outages.[44] Long term, there is a chance these gases may leak and not stay buried. And the vast infrastructure of carbon dioxide pipelines that is required can also disrupt lands and waters. Troublingly, in the oil and gas industry, CCUS almost always means pumping exhaust $CO_2$ underground as a way to push out more oil and gas from aging fields, which will bring more emissions. This technique is called "enhanced oil recovery."

There are serious limits to scale for both CDR and CCUS schemes. There is a finite amount of geologic storage sites suitable for long-term $CO_2$ storage, and of long-term carbon storage that can be added to ecosystems. Because of how small our carbon budget now is, carbon removal may well be needed, but most scientists and practitioners recommend reserving CDR and CCUS for the hardest to decarbonize sectors.[45] However, few "net zero by 2050" plans and models have taken this caution seriously.

Worse, some emissions trading, CDR, and CCUS schemes are modern-day land grabs. Many of the forested areas being traded and monetized are where Indigenous peoples have taken care to conserve the natural environment. The lands they occupy and protect are being commodified by nation-states and corporations for emissions trading, cutting Indigenous peoples out of the deals. Globally, some nature-based solutions are "negatively affecting Indigenous Peoples through displacement, restriction to livelihood practices, and cultural impacts," as geographers at Guelph University put it in a 2020 article.[46] So once again Indigenous peoples and their lands are commodified and exploited, and in many cases, they don't benefit.[47]

Perhaps most alarming, Trudeau's as well as the oil industry's visions of net zero oil and gas have not been applied to "scope 3" or "downstream" emissions. Scope 3 includes emissions from the final use of the oil and gas, meaning burning gasoline, propane, or jet fuel, making plastics, and so on. In Canada, scope 3 emissions are 70 to 80 percent of oil and gas emissions.[48] So when governments and oil and gas companies talk about the industry getting to net zero, they are almost always only talking about the 20 to 30 percent of the emissions that come from production (scope 1 and 2) alone, not the products as a whole. Most existing net zero pledges from oil companies do nothing to address this issue. However, oil and gas companies have been using net zero rhetoric to claim they have effectively solved, or are close to solving, their role in the climate problem. Almost all the big multinational oil companies have announced commitments to become "carbon neutral" or "net zero" by 2050 or earlier. As of 2021, four of the six largest Canadian oil and gas companies had made such a commitment.

Having even these limited carbon-neutral targets set for 2030 or 2050 conveniently allows continued business as usual for now and even the expansion of oil and gas production and the associated increases in emissions. For this, the oil industry gets lauded as a good corporate citizen and even a climate champion that is ushering in our low-carbon future. But oil and gas, and associated byproducts like plastics, will continue to be more and more abundant in this scenario. We cannot let this trend continue; we must instead reduce actual emissions to very low levels in order to have a hope of limiting global heating to 1.5°C. The convoluted and unreliable arithmetic of "net zero" is a recipe for disaster.

If we look ahead another ten or twenty years, this latest net zero phase of climate delayism is likely to lead to a terrifying place. The solutions behind government and corporate net zero plans are mostly market-based and rely on creating new opportunities for private investors that will facilitate further inequality. They act as a shock absorber for public concern on climate change by appearing ambitious, but in practice leading to dangerous levels of heating that will be disproportionately felt by Global South countries and in Indigenous communities. Even very needed climate solutions will duplicate current inequalities if a market-dominated net zero paradigm steers how they are distributed. We can easily imagine a future where renewable energy is abundantly available for those who can pay for it (to do unnecessary things like mine bitcoin, maintain climate-controlled mansions, and drive luxury electric SUVs), while others struggle to access energy for basic needs, renewable energy jobs are precarious and underpaid, and the associated mining expansion leaves Indigenous communities around the world dealing with pollution and human rights violations.

The dominant net zero paradigm, in combination with governments' refusals to keep promises for international climate finance or accept climate refugees, sets us on a path of eco-apartheid where the wealthiest can pay to adapt to or avoid impacts and everyone else is left behind.[49] There is, unfortunately, enormous capital ready to profit from this slow and inequitable vision for transition. At COP26 the Glasgow Financial Alliance for Net Zero (GFANZ) was announced, an alliance of banks and corporations with combined assets of $130 trillion. The Trudeau Liberals, and their ideological supporters like Mark Carney, the UN Special Envoy on Climate Action and Finance and former Bank of Canada governor, have been able to present this vision as the most pragmatic, achievable path because there's a massive pool of capital keen to make it happen and it does not disrupt our current social relations. But this does not matter if the future it will deliver is not livable.

We need to take control and chart a different path. There have been many inspiring moments of transformative change throughout the years, showing us the possibilities for better. When tree planters and Indigenous peoples and settler environmentalists worked together in the 1990s on bettering environmental and labour practices in the logging industry, they showed the enormous potential of mass solidarity. When Indigenous peoples, researchers, and media makers exposed the destruction happening in the oil sands in the 2000s–2010s, they showed how people can work together to bring important stories to light and change the dominant narrative. And as people in the 2020s discuss Land Back and how to share lands to live up to the Treaty relationships, we are seeing the seeds of how our societies can transform. When the environmental and Indigenous rights movements, which came to prominence in the 1960s, join together along with workers, students, parents, and anyone and everyone else, we see ways to become unstuck from feelings of hopelessness. We see ways to transform our societies without waiting for the politicians—progressive or otherwise—who have let us down for so long.

# 3.
# A JUST FOSSIL FUEL PHASE-OUT

The Trudeau Liberals were elected for a third time in 2021 to a minority government, without a plan to reduce fossil fuel production. This time, they did feel enough pressure to make some promises that *sounded* like a plan. They promised to cap emissions in the fossil fuel sector and drive down Canada's emissions to net zero in 2050.[1] These promises came after the Liberals had bought the Trans Mountain Expansion pipeline in 2018 and increased subsidies to "bail out" the oil and gas sector during the COVID-19 pandemic. With the new promises around the 2021 election, the fossil fuel industry got what it wanted: good-sounding regulations it could live with, like the net zero by 2050 requirement that is so limited, far off, and full of holes it largely allows business as usual to continue. The choice to cap production emissions—rather than capping production of oil and gas outright—does not account for the 70 to 80 percent of emissions released when Canadian oil and gas is burned, and allows the industry to continue indefinitely through "net zero" promises of carbon sequestration, offsets, and other accounting schemes. It also allows the government to continue to approve, back, and invest in pipeline projects like Trans Mountain Expansion and Coastal GasLink.

These measures run counter to what so many Indigenous leaders and climate scientists have been telling us: that we must ditch the fossil fuel economy. Indeed, the International Energy Agency, an intergovernmental body equivalent to the World Health Organization, confirmed in 2021 that, to have a decent shot at limiting global heating to 1.5°C, no new fossil fuel extraction projects should be built and existing ones need to be phased out rapidly.[2] The agency's new scenario considered only the overall global climate budget; accounting for Canada's "fair share" of a global fossil fuel extraction phase-out—given its wealth and historic responsibility for the crisis—would mean phasing out all fossil fuel production by the early 2030s and paying significant international reparations for past fossil fuel emissions.[3] However, the Canadian government and the fossil fuel industry, as of the conclusion of COP26 in Glasgow, were not planning to heed this guidance. Their plan—rather than phasing out fossil fuels and winding down the industry on anywhere near the timeline called for, respecting Indigenous peoples' inherent rights, supporting workers, and investing in needed public goods and services in the process—has instead been to continue flooding the world with high-emissions oil and gas for decades in search of private profit.

## The Canadian Oil and Gas Industry's Plans

The oil and gas industry internationally and here in so-called Canada has consistently said that fossil fuel demand will remain strong long into the future, thus justifying high production levels. This stance is further rationalized by industry claiming the oil and gas economy is essential to Canada's prosperity, and Russia's 2022 invasion of Ukraine has been used as justification for increased production as well. Industry's plans for the future look both similar and different to its past. In Canada, it sees a future in the oil sands. This will look less and less like giant open-pit mines visible from space as the industry shifts to steam-assisted gravity drainage (SAGD), a technique that on average produces *more* greenhouse gas emissions than the mines, but that looks less destructive. We urgently need, instead, to stop new project approvals and quickly phase out existing oil sands projects.

But it is increasingly with fracking shale deposits in northern British Columbia and Alberta—mostly for gas, but for some oil as well—that industry sees its growth. There are plans to expand fracking in British Columbia and Alberta to export most of this material as liquefied natural gas (LNG), with gas production expected to nearly double in the 2020s to 26 billion cubic feet per day by 2030. This puts Canada second in the world only to the United States for new gas production plans over the next decade.[4] Gas has not often been a focus of climate politics in Canada, but that has been changing with the Wet'suwet'en-led fight to stop Coastal GasLink and the many LNG terminals proposed for British Columbia's coast. And many Indigenous peoples have been

opposing gas development, for example, on the east coast in Elsipogtog
in 2013 and more recently around the Alton Gas storage facility, both
in Mi'kmaq territory. Lesser known in the rest of Canada, a grassroots
movement sprang up in Quebec against fossil gas fracking. It has won
major victories, including a ban on fracking on the island of Anticosti in
the St. Lawrence in the 2010s that, in 2021, solidified into a permanent
ban on all new oil and gas development in Quebec. Fracking bans have
also been won in Nova Scotia and on Prince Edward Island.

Gas has long been falsely marketed as a "bridge fuel" that produces
lower emissions than oil and coal when burned and is therefore part
of a climate-safe transition plan. This ignores the realities of our global
carbon budget, the decades of lock-in associated with each piece of gas
infrastructure, as well as the high levels of methane released across the
gas supply chain that can make gas just as GHG emitting as other fossil
fuels.

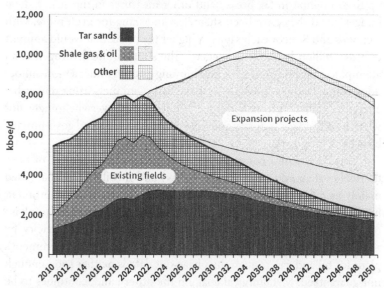

**Projected Canadian oil and gas production to 2050,
by type, and developed vs. new projects**

Source: Dale Marshall, David Tong, Kelly Trout, *Canada's Big Oil Reality Check: Assessing the
Climate Plans of Canadian Oil and Gas Producers* (Environmental Defence Canada and Oil
Change International, 2021), 17, priceofoil.org. Oil Change International calculations from
Rystad Energy UCube data (September 2021); kboe/d = thousands of barrels of oil equivalent
per day.

A newer part of industry's plan to win social licence for expanded gas production is by converting it to hydrogen and making far-off, not-yet-proven promises to use carbon capture, utilization, and storage to bury the emissions. Industry often refers to this process as "blue hydrogen." In northern Alberta and British Columbia especially, there are massive plans to build hydrogen plants and infrastructure in order to create new and longer-lasting markets for expanded fracked gas production. Promising in some cases to pair this expansion with carbon capture, utilization, and storage (see chapter two), industry calls it "carbon neutral" or "net zero," even though the technology required is far from ready or economically feasible.

As of 2021, only five commercial-scale facilities in the world had CCUS technology that was not linked to enhanced oil recovery.[5] All of them have been propped up by massive subsidies and have had trouble meeting their claimed capacity for capturing $CO_2$. Nonetheless, the industry has attracted new subsidies and favourable regulatory frameworks from governments to back and legitimize these plans—including up to $10 billion in tax breaks and other measures in the 2022 federal budget.[6] And they have been shameless in asking for even more, with Cenovus and Suncor each saying a higher tax break that would amount to a $50 billion handout is needed.[7] The shift to blue hydrogen is an attempt to continue producing and burning gas, but under a more innocuous name. It is also possible to make hydrogen fuels using renewable energies—called "green hydrogen"—but that is overwhelmingly not what the oil and gas industry and the provincial and federal governments have been proposing.

Canada is still a consumer of coal and a major exporter of both thermal coal (used mainly for electricity) and metallurgical coal (used mainly as a fuel and reactant to make steel). The federal government has banned new thermal coal mining expansions and pledged to phase out thermal coal exports and almost all coal-powered electricity by 2030. This marks a significantly different approach to the government's plans for oil and gas, though the phase-out date is neither soon enough nor guaranteed. Metallurgical coal, on the other hand, continues to be mined and exported with no wind-down in sight.* In this chapter, we

---

* Metallurgical coal, until recently, did not have a substitute in the steel-making process, and has been treated differently than thermal coal by many climate activists. This has been changing in the 2010s and 2020s.

mainly focus on the oil and gas industry, as it represents a much larger share of our fossil fuel production by emissions than coal and has a stronger political influence, and because the federal and most provincial governments are still highly supportive of the oil and gas industry. We will, however, sometimes refer to fossil fuels more broadly to include coal. As of 2022, active coal mines are located in British Columbia, Alberta, and Saskatchewan.[8] The oil and gas industry continues to take a number of forms, like the infamous oil sands, giant oil rigs offshore of Newfoundland and Nova Scotia, conventional oil rigs in the Prairies, and fossil gas and oil fracking operations, especially from Saskatchewan to northwestern British Columbia.

In the scenario described—where despite "net zero" promises, oil and gas companies keep expanding production—Canada would blow past even its unambitious emissions targets set out in the Paris Agreement and would continue to violate Indigenous rights. The words the industry will use in the future, we expect, will be slightly different than in the past, fine-tuned to be at the leading edge of new denialism. For example, rather than dismiss the need for an energy transition out-right, we increasingly see industry PR, along with friendly think tanks and media, try to convince the world that this transition is far off, with many years of fossil fuel use to come. "The transition won't happen over-night," politicians repeatedly tell us, trying to lower expectations. At the same time, we're also told that oil and gas companies are already playing a central role in the transition with false solutions like CCUS and blue hydrogen. We need to see through this smokescreen.

Thankfully environmental researchers and organizations are doing that work, along with a small but growing number of climate journalists. And sometimes branches of the government call other parts out. After the Liberals announced their 2030 Emissions Reduction Plan in 2022, with detailed modelling but little in the way of new policies, Canada's Commissioner of the Environment and Sustainable Development wrote that the 2030 emissions reductions targets may not be reached, due in part to unrealistic assumptions about blue hydrogen plans.[9] The federal plan was also criticized by many academics and environmental groups for an over-reliance on CCUS while allowing oil and gas production to increase.[10]

The task of taking on the oil and gas industry and charting a differ-ent course is a monumental one because the industry is so politically powerful, with influence in every corner of society, and it is fighting for its very existence. We need to seriously strategize and plan in order to take on the industry and wind it down on a timeline in line with Canada's fair share of 1.5°C, with worker- and Indigenous-led transition plans that do not leave anyone behind. To that end, we should ask ourselves: What will get us moving in that direction, and what will lead us further into eco-apartheid through the new climate denialism of delay? In this chapter and the next two, we look at big changes, often at the level of policy or public investments, that could transform society towards a decolonial just transition; then, in the last two chapters, we will discuss strategies to make it happen.

## Supply-Side Solutions

Under neoliberalism, we have long been told by many politicians and economists that emissions reductions will be driven by individuals changing our behaviours and reducing our consumption of fossil fuels. The role of government, in this approach, is to lightly incentivize individual behaviour changes through policies like carbon taxes, rebates for changing light bulbs or buying electric vehicles, advertising campaigns to buy local, and tax incentives for energy efficiency upgrades. But for many of us who, for example, have to drive to work to make a living and whose food options are limited to low-priced supermarket goods, sourced from around the world and grown with massive petrochemical inputs, there are only a few behaviours we can change in the course of getting by, and many we can't. These approaches to climate change mitigation can be described as "individual demand side," as they focus on reducing demand for oil and gas and see individual consumers as the sites of change.

This individualized demand-side approach, which has dominated for decades, is severely limited. At its worst, it promises that we can make big changes without directly transforming how we produce goods and provide services in our society; by blaming individuals for collective inaction, it can be demoralizing. When its logic is taken to the extreme, bigger transformative changes—such as opposing a new oil pipeline or highway and advocating instead for alternatives like better public transit—are seen as illegitimate. Proponents of individualized demand-side approaches, including the fossil fuel industry, argue that you can't fight for big changes while also being reliant on fossil fuels in the present, using plastics, heating your home with fossil gas, or fuelling your car with gasoline. Responsibility for climate action is put on the consumer choices of each person, as if one individual can change the course of capitalist industrial development by changing their habits while all else remains the same. Despite its absurdity, politicians, business leaders, academics, and even environmental groups have for years been tightly focused on these individualized demand-side actions. Through it all, the fossil fuel industry has grown and grown, with emissions going up and an ever-growing list of Indigenous rights violations.

There is certainly a role for collective demand-side solutions, especially those that do not rely on market logic, such as massive programs to retrofit buildings to use less energy or coordinated boycotts of certain harmful products and services. Organized demand-side solutions on a scale big enough to have a significant impact have been written about widely, and we cover our vision for these strategies next chapter. But we need to beware of individualized demand-side approaches that don't challenge the status quo, such as telling people to take public transit but providing no good options, as these have only limited effect and can even work as a form of climate delayism by promising change and not delivering. And as geographer Mazen Labban points out, because the financial sector and its convoluted financial instruments, like derivatives, are taking up a growing share of economic activity (a process called "financialization"), changes in demand for fossil fuels have a much less direct effect on supply than basic economics lessons would suggest.[11] Though reducing fossil fuel demand is a net good, markets sometimes produce counterintuitive results; for example, oil demand decreases might be accompanied, because of the complex nature of financialization and other factors in global markets, by oil price increases while production (oil and gas supply) stays constant or even goes up. Demand-side approaches, especially individual-focused ones, are simply insufficient on their own.

It's time to consider bolder proposals that have the ability to drastically reduce emissions on a timeline consistent with avoiding the worst impacts of the climate crisis. Thankfully, social movements, progressive policy communities, and think tanks have been looking at how best to implement supply-side policies targeting the oil and gas industries that produce and profit from the sale and continued demand for oil and gas around the world. Put another way, supply-side climate policy echoes the well-known rallying cry about fossil fuels in the climate justice movement: "Keep it in the ground!" It turns out that the movements are right: by constricting the supply of fossil fuels, changes to consumption patterns are also achieved. New fossil fuel extraction projects and the infrastructure they come with tend to last twenty to forty or more years and locks in higher levels of supply by creating high legal, economic, and political costs to retiring the projects early.[12] By banning the building of new fossil fuel projects, we can avoid allowing fossil fuels to flood the market and drive continued demand. By signalling an end to fossil fuel expansion and towards its managed decline, we also spur on alternatives. "Restrictive supply-side policy instruments (targeting fossil fuels) have numerous characteristic economic and political advantages over otherwise similar restrictive demand-side instruments (targeting greenhouse gases)," political theorist Fergus Green and economist Richard Denniss write in *Climatic Change* journal. They argue we should be "cutting with both arms of the scissors," meaning, to cut emissions we should use both demand-side and long-neglected supply-side approaches.[13]

A supply-side approach has the added benefit of diminishing the power and influence of fossil fuel corporations in our societies by regulating, taxing, and constraining the industry. These strategies send a clear signal to fossil fuel companies that they will be of decreasing importance in charting a path for our future and materially reduce their market value. Focusing on the supply side can move us in the direction we need to go at the pace required to avert global catastrophe. And such policies can be combined with visionary demand-side approaches. Importantly, supply-side policies, which constrain the production and, thus, extraction of fossil fuels are consistent with the many Indigenous-led land defence movements demanding materials be kept in the ground.

We know we need to decarbonize quickly, and we know that phasing out fossil fuels is key to that decarbonization process, and to wresting our economies and societies back from the influence of fossil fuel interests. We should work towards five interrelated goals: diminishing the power of the industry, respecting Indigenous rights, quickly reducing production of fossil fuels, shifting revenue streams in the final years of industry away from private investors and towards the public good, and protecting workers and communities who rely on the industry during the transition.

## Regulatory First Steps

Winding down the fossil fuel industries will be hard work; it will require us to push governments to do things that are outside of traditional mainstream approaches to climate action in so-called Canada. There are regulatory first steps that would weaken the power and influence of the industry and contribute to immediate reductions in GHGs. These regulatory first steps, which can be taken immediately, will focus the public's attention on making the polluters pay and prepare the ground for potentially more significant intervention to come. We draw inspiration from a document published by Oil Change International and The Next System Project at the start of the pandemic in April 2020 outlining measures to wind down the fossil fuel industry in the United States.[14] These first steps would signal to industry that there will be no more expansion, and would lower the value of fossil fuel assets by making it more expensive for companies to operate. Once fossil fuel assets have been devalued and the public's attention is on the legacy of pollution and rights violations, the industry will be on its heels.

Ending subsidies to fossil fuel corporations is "low-hanging fruit" that will speed up the shuttering of operations. It's also very politically popular and would immediately free up public money to support just transition efforts. On average in 2017 to 2020, Canada's provincial and federal governments provided $10 billion in direct subsidies to the oil and gas industry and $14 billion more in support through preferential government-backed finance.[15] These figures include tax breaks, drilling incentives, handouts to specific projects like the Trans Mountain Expansion pipeline, and royalty rebates, among other measures—and are still an underestimate, since data is not fully publicly available and other subsidy data like tariff exemptions and clean-up costs that companies offload onto the public are missing entirely. In addition, provincial and federal governments spend considerable time and resources promoting the industry, including promoting exports through the federal Crown corporation Export Development Canada (EDC). These subsidies and marketing activities should be cancelled immediately, as a simple and straightforward action towards climate justice.

Perhaps because subsidies are one of the most visible forms of Canada propping up the fossil fuel industry, the Trudeau Liberals have been pushed to make concrete promises on them. Prior to the 2021 election, the Liberal's party line was actually that no more "inefficient" fossil fuel subsidies remained. The Harper Conservatives first promised to phase out "inefficient" fossil fuel subsidies, along with other G20 countries, in 2009. Despite every subsequent Canadian government reiterating this promise, very few subsidies have been rolled back and new ones have been introduced. The Liberals long denied that payments to industry to clean up wells or reduce emissions are subsidies, even though they have the same effect of lowering companies' costs and have been shown not to increase the rate of clean-up, and the government also denies that tax breaks constitute subsidies.[16] The Liberals were pushed to change tack slightly in 2021 only because of public pressure and the Bloc Québécois and NDP pushing hard together, as the two parties' own positions on the issue were aligned and much stronger than that of the Liberals.

The Liberals made two new promises. At COP26 in Glasgow in late 2021, Canada and other nations pledged to end direct international public finance for "unabated" coal, oil, and gas—meaning projects without something like carbon capture to "abate" emissions—by the end of 2022 and to prioritize clean energy finance. And in the 2021 election, the Liberal government pledged to phase out other fossil fuel subsidies by 2023. But there is great risk of these promises being broken or implemented in half-steps by using slippery definitions, as we have already seen with the new tax break subsidies for oil and gas using CCUS that the Liberals introduced in 2022. Worse, the promises do not yet cover all of our subsidies. Neither commitment (on international public finance or on other fossil fuel subsidies) explicitly covers Canada's *domestic* public finance for fossil fuels, even though it is a subsidy. Domestic public finance averaged somewhere between $5 billion and $12 billion a year in 2017 to 2020, according to research by our co-author Bronwen Tucker for Oil Change International.[17] The total will depend on how the federal government decides to define "domestic" versus "international" in order to uphold its pledge to end international public finance—a time-wasting distinction given all of this support must end. And a large amount of fossil fuel handouts remain at provincial levels, especially in British Columbia, Alberta, and Saskatchewan, with little end in sight.

Alongside ending fossil fuel subsidies to lower the artificially high value of fossil fuel assets, early decisive action would also include regulations to the same effect. First lands, especially ecologically sensitive ones, could be set aside and made unavailable for development. Governments could also set immediate and declining production caps, which would signal the sunsetting of fossil fuels and give companies some time to plan accordingly, all while making it more expensive to pollute. Enforcing and strengthening currently paper-thin regulatory systems would accelerate this process. Interestingly enough, there was a recent example of regulated production caps when the Alberta NDP responded to the oil price crash in fall 2018 by mandating that oil companies cut production. This limit lasted from 2019 through 2020 as a way to keep the price of oil high, rather than letting it bottom out. It was not a climate measure, even if it may have reduced oil production temporarily, but it does show us that governments have the power to mandate production declines if they want, even in the face of backlash from some corners of private industry.[18]

Holding corporations responsible for cleaning up their tailings waste and abandoned wells and for site reclamation should also be an early policy priority, since it can drive a wedge in the industry's claims to be a net positive benefit for society. And it will create good, green jobs where they are needed as fossil fuel producing regions begin to transition. Leaked documents from the Alberta Energy Regulator in 2018 reported that liabilities in Alberta's oil and gas industry alone added up to $260 billion, compared to the regulator's earlier public estimates of $58 billion.[19] Yet because the industry has had so much sway in designing the regulatory program, the province, as of 2021, only holds $1.5 billion in securities from the companies. It is clear that the oil and gas industry is a major source of risk to the public. The Alberta Liabilities Disclosure Project, which is advocating to hold industry to account for cleaning up its mess, is calling for the full cost of clean-up to be collected during the course of a project's life, installing a reclamation levy across the industry, and for legal reform to prevent executives and investors from walking away from their clean-up responsibilities by declaring bankruptcy.[20] If we held individual oil and gas executives and investors responsible for environmental liabilities, we could see a very different attitude to the problem and access to much more funding.

Implementing and enforcing the United Nations Declaration on the Rights of Indigenous Peoples, as it was intended at the United Nations level, is a key regulatory first step. But to be truly effective, Canada has to concede that rights such as self-determination and free, prior, and informed consent are not subordinate to the Canadian Constitution or Canadian sovereignty. If genuine free, informed, and prior consent processes were enforced, a number of large pipeline projects such as Coastal

GasLink and Trans Mountain Expansion would have to be abandoned. This alone would not strand all oil and gas assets, but it would signal to investors that the industry is not in an expansion phase, a calculation likely to decrease the companies' value.

Finally, there is also value in creating new ministries, agencies, and Crown corporations tasked with facilitating and coordinating the wind-down of the fossil fuel industry and the ramping up of renewable energies. As Seth Klein points out in *A Good War*, in the Second World War, an extensive list of Crown corporations was created to manage the industrial planning and production aspects of the war effort. We need to be similarly ambitious in industrial policy, though not to the aims of war and ever-increasing mechanization, but to peace, just transition, and repair of lands and life.[21] In March 2022, activists and politicians showed what this could look like, holding a press conference with announcements from fictional agencies, like the Ministry of Just Transition, the Land Back Secretariat, and the Clean Transit Without Delay Commission, all reporting back on the amazing transformations that had taken place—or could have—by 2025.[22]

## Public Ownership or Regulation?

To phase out the oil and gas industry, there are advantages to putting the industry under public ownership and also to regulating it without public ownership. The process for putting a company or whole industry under public ownership is sometimes called "nationalization," meaning the federal or provincial government takes possession and owns it. But given the colonial nature of that specific process, we prefer to use "public ownership" or "socialization," as we see the need for Indigenous leadership and rights assertion in this process. We also see the need for reparations to Indigenous Nations as part of winding down the fossil fuel industry, to account for the long-term theft and despoiling of lands. This funding could come from companies and governments, and it would need to happen in parallel with regulations and, if deemed appropriate, a public ownership process. Pathways exist for both public ownership and regulation, but obstacles become apparent as we work through each option. The bottom line is that both paths require strong political will and organized social movements that will hold governments accountable for a just phase-out. Any progress is at risk of being overturned or co-opted by industry and government alike, prolonging the colonial fossil fuel era. We will focus on the question of how to build the required political pressure in chapters six and seven.

Taking the regulation approach, governments would leave the industry in private hands and mandate declines in oil and gas production, thereby winding down the industry on a set timeline. Trudeau's oil and gas emissions cap would need to be transformed into a production cap. Mandated declines would be immediate, not targets four or eight or thirty years out, as governments have thus far been setting. Governments in Canada are only in power for four years at a time and elections are often called before the time is out, so changes need to happen when governments have the power to make them. Otherwise oil and gas industry executives and their investors will delay action and lock in new infrastructure, making it more difficult for successive governments to reverse the trajectory.

There is also the thorny question of which governments should act. Indigenous governments, which have not given up their rights to the land, must be involved in a way that recognizes their jurisdiction and authority (we'll return to this theme in the next section). Within the colonial framework, provincial governments have jurisdiction over natural resources and regulate their industries, with the federal government having some limited powers. Achieving coordination across jurisdictions, some of which are contested, will be difficult. We have seen, though, that the federal government can play a role by instituting backstop policies and regulations while allowing provinces to chart their own courses. The legal precedent for this was solidified by Trudeau's carbon pricing regime, which withstood a Supreme Court challenge launched by Saskatchewan. Similarly the federal coal phase-out and methane regulations show that the federal government can set a minimum bar and provide policy direction on GHGs related to resource extraction.

A drawback of leaving the industry in private hands and mandating a phase-out is that it is very difficult to imagine this approach working at the required scale. Private, for-profit oil and gas companies, as long as they exist, will fiercely oppose overall declines in production, and will do what they can to get governments to abandon such a direction. For example, they might coordinate "capital strikes," meaning to pull investment from an area until they get their way, which can cause economic chaos in our capitalist economy. While we want the industry to decline, we need it to happen in a managed way that protects workers and communities while we transition. In other words, a planned economic transition is required.

Another drawback of trying to regulate the private industry out of existence is that government regulatory agencies are already significantly "captured" by industry and so work in their interests, this being the result of decades of concerted industry efforts. There has been a revolving door between oil companies and regulatory agencies, with staff going back and forth. There are dogged lobbying efforts, media offensives in collaboration with think tanks and academic allies, and funding of political parties. The private industry has proven highly effective at making sure oppositional politicians who propose tough regulations don't get in the way of their profits. Former Alberta Liberal leader Kevin Taft described an illuminating episode in his book *Oil's Deep State: How the Petroleum Industry Undermines Democracy and Stops Action on Global Warming—in Alberta, and in Ottawa*.[23] Taft was known for trying to take on the oil industry in Alberta as leader of the provincial opposition; during that time a powerful individual in the industry told him if he kept at it, his political career would be over. According to Taft, he kept going at the industry and was never elected again.

Private oil companies and the investors who own them have also been resistant to even modest measures to uphold Indigenous rights. At the Enbridge annual general meeting in 2017 in Calgary, for example, a shareholder motion was put forward to have the company "identify and address social and environmental risks, including Indigenous rights risks, when reviewing potential acquisitions." This modest proposal was defeated by shareholders, with the company's board saying the pipeline giant, which was at the time building the Dakota Access Pipeline despite a huge Indigenous resistance effort at Standing Rock, was doing enough already. Indigenous peoples resisting the pipeline were met with a truly massive police and security presence; among other tactics, they were blasted with fire hoses and attacked by dogs.[24]

The socialization option would involve bringing the fossil fuel industry in Canada under public ownership, rapidly but humanely shutting it down, and developing alternatives. Bringing fossil fuel companies under public ownership would take away the profit motive and put them under the control of governments. If done well, this approach would allow for much more certainty in the outcomes for emissions trajectories as well as for workers and communities. And one of the largest industries in the country and fiercest opponents of climate action and Indigenous rights assertion would no longer be spending millions convincing the public and politicians to allow it to continue its destructive ways. All the lobbying, all the ads for gas stations, all the industry sponsorship of think tanks and research centres could be over, a relic of the past.

There would likely still be reactionary elements pushing to make the industry private again. After all, business-friendly pundits, academics, and think tanks have pushed for the privatization of the healthcare system since it was socialized in the 1960s, with some success. But these elements wouldn't have the same financial backing from multi-billion-dollar oil and gas companies as they do today. This would be a huge advantage for making progress on climate and Indigenous rights. If the regulatory first steps we propose here are done first, it would also be a much less expensive option than it might seem at first glance. Fossil fuel companies' valuations are artificially high because they have existed in a regulatory context that gives them false advantages at every turn. Removing these advantages, plus adequately factoring in the existential threats to this sector from global decarbonization, should make negotiating the purchase of these companies not prohibitive. Expropriation of oil and gas assets without compensation to the companies, in service of the public good, is another, though more complicated, possibility.

## Shared Jurisdiction and Dual Governance

Let's not fool ourselves that government ownership of the oil industry is itself sufficient to phase out fossil fuels and do it in good ways. We live in a context of powerful colonial governments looking to generate revenue for themselves, ideally without raising taxes, that have shown almost no ability to respect Indigenous sovereignty. Oil operations being run by colonial governments like the provinces or Canada could be just as bad violators of Indigenous rights as the private companies.

A just version of socialization would require, at the least, shared control of the fossil fuel industry between Indigenous peoples and settlers. This may seem difficult to imagine because the history and present of Canada is one of denial of sovereignty and Indigenous rights. But, as Cree and Saulteaux scholar Gina Starblanket reminds us, the historic treaties already lay out the framework within which dual governance and shared control might be accomplished. "Far from a 'sale' of land," Starblanket writes, "treaties are regarded by Indigenous peoples as land-use frameworks, which generally involve the establishment of separate governments and jurisdictions in distinct spaces, and dual governance and jurisdiction in shared spaces and matters of mutual concern."[25] We can use these frameworks to propose bold new initiatives and respectful land uses.

Instead of the colonial federal and provincial governments headquartered in Ottawa, Regina, Edmonton, or St. John's calling all the shots about the industry, we need powerful movements of people, organizations, and pressure groups agitating for genuinely shared jurisdiction and management with Indigenous Nations, as per the Treaty relationship (or in places like much of British Columbia, where there is no Treaty relationship, per genuine government-to-government negotiation). We cannot keep repeating the sham of what colonial governments call "co-management," but which really means the settler state maintains full control. Control must actually be shared where Nations overlap spatially and have mutual concern, but in other cases, full Indigenous authority and control makes sense. In other words, Canada does not get to assume it has an interest over the entirety of what is currently called Canada.

In terms of the settler side of governance, we can imagine ways of distributing control away from the existing centralized governments and towards local communities. For example, instead of settler governments imposing their will without involving local communities, regional boards and committees could be set up to help direct publicly owned oil and gas operations, leading the wind-down process. These could be made up not of wealthy elites but of rank-and-file workers and local community residents.

To make it concrete, let's imagine public ownership in Treaty 6, which stretches through central Alberta and Saskatchewan, where much of the oil sands are located. As discussed in chapter one, Indigenous Nations in Treaty 6 never ceded their lands, not to mention sub-surface rights, to settlers, and so the future extraction of oil and gas there badly needs to be renegotiated. This negotiation could happen in government-to-government processes between the Treaty 6 and Métis Nations and the governments of Alberta, Saskatchewan, and Canada.* Settler govern-

---

* Land rights in the colonial system are now largely managed by the provinces through the Natural Resources Transfer Acts—with the provinces acting as "the Crown"—but the federal government continues to claim authority over some lands.

ments could develop ways to make their side of the process open and democratic, directed more by local people rather than by bureaucrats and politicians in provincial and federal legislatures.

Land-use frameworks for (declining) oil and gas extraction would need to implement each Nation's separate jurisdiction, laws, and systems in their own territories. "Crown lands" would not be automatically considered Canada's jurisdiction; instead, as part of Indigenous territories, their land uses would be directed by the relevant Nations. This approach will also require Indigenous peoples to more fully develop and implement their own constitutions, environmental assessment practices, and development protocols, in line with their governance, laws, and traditions. Hundreds of years of imposition of colonial laws and governance have eroded the practice of self-governance. When Canada finally steps back, Indigenous jurisdiction can flourish once again.

Where territories and interests overlap, all parties would come equally to the table. Instead of Alberta coming to the Treaty 6 and Métis Nations with projects already fully planned out (at the direction of oil and gas companies), all parties would engage in planning together, and Indigenous Nations would be given the time needed to discuss plans internally. Too often, companies and colonial governments come to First Nations last, pushing them to sign agreements that supposedly give approval to a project, without time for engagement with the Nation— including through traditional governance systems, not just the Band Council system imposed by the Indian Act. The approach here would be in line with UNDRIP, and especially mindful of several key articles. Article 19 requires "free, prior and informed consent" on legislative or administrative measures"; article 20, the right to maintain and develop Indigenous "political, economic and social systems and institutions" and engage in subsistence, traditional, and other economic activities. Article 32 states that "Indigenous peoples have the right to determine and develop priorities and strategies for the development or use of their lands or territories and other resources." These processes would also need to be in line with the local Treaty relationships based on trust, respect, and good faith.

From such an engagement, plans could emerge for a rapid, well-planned winding down of fossil fuel extraction to near-zero on a timeline consistent with the scale of the crisis. Profits from the sale of fossil fuels between now and then could go to help First Nations in the region build new renewable and energy-efficient infrastructure and green economy projects as well as meet other pressing community needs. And the proceeds could potentially pay for infrastructure and services to reduce emissions and improve life in rural areas of Treaty 6 as well as its cities, including Edmonton, Saskatoon, and Prince Albert. Some profits would also need to be used to properly shut down oil and gas infrastructure and remediate and reclaim lands from a shrinking industry.

While we cannot presume to know what Indigenous-settler sharing of jurisdiction and management of fossil fuels would look like and what decisions would be made, we are confident that fossil fuels can be phased out in a way that upholds Indigenous sovereignty and jurisdiction and that direct the revenues from a shrinking industry to life- and land-affirming projects and goals.

## Supporting Fossil Fuel Workers and Communities in the Transition

If this transition is to be just, phasing out fossil fuels will require significant investments in workers and their communities. The concept of a just transition has deep roots in the labour movement, making the well-being of workers a priority. So far, this book has focused mainly on Indigenous Nations and their sovereignty and well-being. That is because Indigenous peoples have long been the communities most neglected and mistreated by colonial governments and resource industries. But many working-class settler, newcomer, and Indigenous peoples alike are precariously positioned in resource extraction. Women and Indigenous and racialized workers are less likely to be employed directly in the oil and gas sector, but are overrepresented in low-paid service sectors, like retail and social services, that are essential to the well-being of fossil fuel communities and could, without careful planning, be even more affected by a production phase-out.[26] And while jobs directly in the oil and gas sector may be high paying, many of them are non-unionized and treat workers as disposable. Resource companies regularly lay off workers, reduce work hours, and cut benefits when market conditions change or operations shrink or close down. As Parkland Institute research has shown, from its peak employment of more than 200,000 workers in 2014, the Canadian oil and gas industry had shed more than one-quarter (53,000) of its workers by 2019, even while oil production continued to grow.[27] It is essential in a just transition away from fossil fuels that all workers, including fossil fuel workers, are supported to live dignified lives where their needs are met.

Fossil fuel workers have skills that are required to make the transition to renewable energy and a low-carbon economy. Groups like Iron & Earth are organizing these workers, providing training to repurpose their skills to work in the renewable industries. This kind of initiative will be needed to ensure that fossil fuel workers do not continue to vocally oppose the winding down of the fossil fuel industry. After all, many fossil fuel workers have legitimate fears that they will lose their jobs and struggle to maintain a certain quality of life. While the fossil fuel industry is keen to play up the fears of its workforce to its own ends, there is also truth to workers' concerns. This is not because an energy transition can't look out for everyone, fossil fuel workers included, but because previous energy and economic shifts have left many workers behind. The history of capitalism in so-called Canada is full of instances of workers being left to fend for themselves once corporate profits dry up, from the cod fisheries on the east coast to manufacturing in Ontario. These precedents are a function of how governments have not been there to support workers, preferring to support executives and shareholders. In addition, neoliberal governments have whittled away many of the social safety nets such as employment insurance and income supports, as well as public programs such as social housing, making life and work more precarious for nearly everyone.

In Alberta and Ontario, former coal workers are familiar with governments mandating energy transitions without doing adequate economic planning or giving enough financial support to workers. The coal phase-out in Ontario from 2003 to 2014, under the provincial Liberals, was done with hardly any input from workers, and communities were essentially abandoned by the provincial government. In Alberta, the NDP legislated the end of burning thermal coal for electricity generation in the province. As the Parkland Institute has pointed out, affected workers and communities received between $54 million and $214 million in support through a number of programs: topping up employment insurance for laid-off employees, retraining for new jobs (but not actively connecting laid-off workers with those new jobs), providing an early retirement option for people at least fifty-three years of age who had worked for at least ten years for their employer, a moving allowance of $5,000 for people to move to new communities to follow job opportunities, and broader economic development funds for new infrastructure and renewable energy projects in the region.[28] But in contrast, companies burning coal to generate electricity were paid out a total of $1.36 billion dollars, at least nineteen times the amount provided to workers. Though they had received some support, workers were understandably upset.

According to Alberta Federation of Labour president Gil McGowan, whom a couple of us interviewed for *Briarpatch* magazine in 2018, the supports fell short, since workers were not connected with a new job and their communities were not supported to truly shift away from fossil fuels.[29] It is unsurprising, then, that inadequate funding for programs to support people affected by shutting down fossil fuel industries fostered resentment. The NDP was not re-elected in 2019 in the ridings with former coal-generating plants. The big corporations benefited the most, and the transition was from one fossil fuel to another, with most of the power plants shifting from coal to gas. And while gas-fired generation can have lower emissions than coal when methane leakage is controlled carefully across the supply chain (never a guarantee), environmental movements were not satisfied, since fossil gas is not nearly low enough to get near to zero emissions. Without a robust, credible program to support workers, made in close consultation with workers, we can expect fierce opposition to a transition off of fossil fuels in these communities.

One key step we need to begin now is organizing to win fossil fuel worker and community support for and leadership towards a just transition away from fossil fuel production. We firmly believe this will not be as hard as media-magnified rumbles of "Wexit" or "I <3 Oil and Gas" bumper stickers would have us believe. Very little effort has yet been put in to counter the status of just transition as a untouchable issue in Alberta and Saskatchewan, which means no alternative economic pathways have been seriously proposed by mainstream media or politicians. In the absence of such an effort, there have been openings for far-right, xenophobic, and white supremacist leaders to get their message across to oil and gas workers and business owners. From the Canadian Yellow Vest Convoy to United We Roll to the "Freedom Convoy," we are witnessing a troubling rise of a politics that scapegoats and puts Black, Indigenous, LGBTQ2S+, and many other communities at risk. We need to counter this trend and to deflate the rising far-right extremism.

In our own experience speaking to, working with, and living in communities where fossil fuel production is the major industry, there are real and growing concerns below the surface. Many workers and communities in the areas most dependent on fossil fuel production are deeply worried about their future, about the health impacts of this industry on their families, about job cuts from automation, and about the viability of the sector amid global trends towards decarbonization. In southern Alberta and Saskatchewan, where oil and gas jobs have been in a precarious position since a major price crash in 2014 that prompted mass layoffs, communities have already experienced how company CEOs and investors dodge local taxes, falsify bankruptcy, and guard their profits, leaving the communities high and dry.

In the coming chapters, we'll discuss some of the good, unionized work that will be ramped up through green infrastructure and economies of care. This work will provide dignified lives for current fossil fuel workers and others labouring in industries that service fossil fuels. Many people are today working in highly exploitative jobs with poor conditions, all across Canada and also around the world. But it does not need to be this way.

# 4.
# GREEN INFRASTRUCTURE FOR ALL

A just transition will not materialize out of thin air when fossil fuels are swapped out for green energy infrastructure. For the energy transition to be just, green, and decolonial, it will also have to undo the theft of land and life that currently underlies the economy of so-called Canada. The energy transition is an opportunity for us to rethink how we relate to one another and to the earth—to learn to live within the earth's bounds while ensuring a good life for all. It is an opportunity to build out infrastructure that is governed by people rather than profit motives, while producing good, unionized, public sector jobs and lowering our use of fossil fuels and other critical minerals by consuming collectively rather than individually.

Imagine for a moment electrical grids that are publicly owned and governed democratically through the overlapping jurisdiction of Indigenous peoples and the public sector. Imagine existing energy utilities working hand in hand with Indigenous Nations to enable renewable energy projects that are community-owned and tied into larger regional grids. Imagine First Nations exercising full authority over their reserve lands and jurisdiction over the full extent of their traditional territories (overlapping, where appropriate, with others jurisdictions, including Canada's). Indigenous Nations would no longer be mired in paternalistic and bureaucratic processes imposed by the federal Crown under the Indian Act to pursue on-reserve clean energy development. Rather, Nations across so-called Canada could create meaningful employment and revenues on their own terms in their own territories—not just reserve lands but the full extent of them! A green energy system that allows for this kind of imagined future would uphold and respect Indigenous rights and sovereignty.

When settler societies refuse to uphold the inherent rights of Indigenous peoples, green energy projects and programs threaten to only continue unjust colonial relationships. Although settler existence in so-called Canada depends on the lands and resources of sovereign Nations and peoples, federal and provincial governments and their institutions have continued to make unilateral decisions about the use of Indigenous lands and to treat Indigenous peoples as clients and dependants. Mainstream green energy transition advocates miss this fundamental injustice, powering ahead with a transition as a mere technical feat. They too often narrowly understand the path to decarbonization as follows: electrify everything and then make the sources of our electricity green. Instead, we need a vision and a plan to create green infrastructure that upholds and recognizes Indigenous sovereignty and redistributes consumption and ownership so that we can all live a good life while respecting ecological limits.

Unfortunately for Akamihk Montana First Nation, their efforts at pursuing green energy for their community have been far from liberating. Located in Treaty 6 territory between Red Deer and Edmonton, the Nation started with a 100-kilowatt rooftop solar installation to power their administration building, a place that serves as a recreation centre, satellite health facility, funeral home, community centre, and seat of government, and is described by band member Vickie Wetchie as "the heart of our Nation." As Wetchie explained to us in an interview, after their first project in 2010, the Nation went on to pursue other energy efficiency and solar projects, both residential and commercial, all the while training their own band members and other Indigenous solar installers across Alberta—over sixty-five in all.

When the Notley government came to power in 2015 and indicated a strong desire to partner with First Nations on renewable energy, Montana First Nation was optimistic about completing the utility-scale solar project that they had already advanced through a feasibility study. Wetchie, through appointment by Montana First Nation leadership, began participating in the Indigenous electricity technical working group, where she sat at tables with the Alberta Utilities Commission and government representatives to work out how Indigenous Nations could be part of the government's new climate leadership plan by producing clean energy.

Yet at the same time, her Nation was experiencing roadblock after roadblock to bringing their solar farm to reality. Multi-level negotiations with the federal and provincial governments and with the Alberta Utilities Commission made it nearly impossible to bring the project online, because these institutions did not recognize Montana's jurisdiction, even on their reserve lands. As Wetchie told us: "It was very difficult to get over that hurdle, to negotiate with them, and very costly. We had legal counsel every time we were in those rooms, and they brought theirs and they don't tell you that they're bringing their lawyers to sit at those tables. And we were ready. We had done our homework. We were ready to have those discussions, we weren't going to let them tell us, 'You are going to do this to get that.' We just weren't going to be bullied in those rooms."[1]

Montana First Nation eventually did prevail and through difficult negotiations were able to make use of the provincial grants that the Notley government had made available. When we spoke to Wetchie in May 2021, the project was nearly online, with the last testing that would enable commissioning just coming to a close. The project will produce 4.8 megawatts that will be sent back to the provincial grid. But this came about only because of an extraordinary commitment from Montana First Nation leadership and the community to fight for the project.

There is not yet a clear pathway for other Nations trying to follow suit, and green energy did not prove to be the promised opportunity for decolonization and self-sufficiency. Federal acts (such as the Indian Act and the First Nations Land Management Act) put up insufferable road blocks to the development of the Nation's solar project on reserve lands, the Alberta NDP government had its own set of timelines and requirements for accessing its green energy project funding, and the Alberta Utilities Commission claimed authority and onerous fees to connect the project to the provincial grid. In these multi-level negotiations, the jurisdiction and authority of Montana First Nation over their lands and their project was consistently misunderstood, Wetchie said.

Many studies have shown that we have the technologies and the know-how to accomplish a green energy transition, right here on the lands known as Canada. Continued innovation and new technologies will help us decarbonize faster and in ways that reduce the monetary, social, and ecological costs of transition. The main obstacles to transition, though, are not technical; they are political. People working in the fossil fuel industries and in the industries that service them and their workers (including construction, hotel and restaurant, and so on) are worried about a future without these jobs. Working-class and poor people experience carbon taxes as costs imposed by an elite that is sacrificing little itself and plowing ahead with carbon-intensive business as usual. Some Indigenous communities are trying to develop green energy projects but are being held back by colonial systems that don't accept their autonomy and jurisdiction.

As long as green energy transitions are seen as an added cost to individuals and society, they will be met with resistance, rather than support and mobilization. And as long as green energy programs continue to deny Indigenous Nations their jurisdiction and authority, they will act as barriers to Indigenous rights. The task before us, then, is to lay out a vision for green infrastructure that is politically irresistible—one that is capable of mobilizing widespread support by making people's lives more affordable, building more resilient communities, decarbonizing at a pace that is in line with a globally equitable pathway to 1.5°C, asserting Indigenous rights, and putting settlers and their governments in good relations with Indigenous peoples in the process.

## Two Principles to Guide Green Energy Infrastructure for All

We see two principles at the heart of a just green energy transition. First, for a just transition, Canada must meet its government-to-government agreements with First Nations, Métis, and Inuit, which will require accepting that its sovereignty and authority are not superior. Instead, Canada must recognize Indigenous laws and sovereignty categorically, and work within and alongside Indigenous jurisdiction. Green energy infrastructure and economies must be planned and implemented in ways that do not endanger the lives, land uses, and livelihood practices of Indigenous peoples, as so excellently laid out in Beaver Lake Cree Nation's court case on the cumulative impacts of oil sands development in their territories. Together with Indigenous peoples, Canada will need to rework relationships of management and responsibility over everything from Crown lands and natural resources to institutions of care and education. As we've discussed in previous chapters, where they apply, we see the implementation of historical treaties as land-use frameworks as the key to recognizing Indigenous jurisdiction and sharing land and authority, whether in winding down the oil and gas industry or winding up green infrastructure.

The second principle at the heart of a just green energy transition is building infrastructure that allows everyone to meet their basic needs while remaining within global ecological limits. The clearest pathway to this goal is taking essential goods and services out of the market and restructuring their provision so they are accessible to all through universal, free, and low-cost public programs. Our basic infrastructure—electricity, housing, transportation—should be operated for the public good rather than private profit, under systems of governance that recognize separate Indigenous jurisdiction in some spaces and overlapping jurisdiction with Indigenous peoples in matters of mutual concern. Restructuring this infrastructure to give everyday people the ability to shift consumption away from individual, private, and high-emissions choices (like the private automobile) and towards lower-impact collective systems (like public transportation) will help us to decarbonize while offering people a higher quality of life. Public sector unionized jobs that provide security and benefits—jobs that are equally there during times of boom and bust—should be the backbone of this strategy.

Canada has relatively high historic greenhouse gas emissions and has profited enormously from the fossil fuel economy that has facilitated these emissions, and so affording the needed atmospheric space for people in the Global South to live a good life means cutting emissions here more quickly. Of course, Canada's wealth and consumption is not well distributed. So a key part of adequately reducing our emissions, distributing our consumption of resources more equitably, and funding this public and collectively consumed green infrastructure is increasing our taxation of the wealthy and corporations. We can build out a series of high-quality collectively consumed goods and services offered at low or no cost to all, including museums, parks, libraries, and other amenities. By doing so, we'll create stronger, more mutually supporting and resilient communities.

## Redistributing Wealth and Emissions

The top 1 percent of people hold almost 26 percent of wealth in Canada, and this wealth hoarding seems only to be increasing. It spiked dramatically during the COVID-19 pandemic; Alex Hemingway at the Canadian Centre for Policy Alternatives estimates the forty-seven billionaires in Canada increased their wealth by $78 billion in 2020 alone.[2] The wealthy are also disproportionately responsible for emissions because of their lavish personal consumption—worldwide, the wealthiest 5 percent con-tribute more than a third of global emissions.[3] So taxing the rich and cor-porations can help lower emissions, distribute them more fairly, and free up money needed to pay for accessible green infrastructure all at once.

The sums involved are staggering, and even moderate taxation could amount to nearly $100 billion a year. Hemingway, for example, estimates that a progressive wealth tax starting at 1 percent on net worth over $10 million could raise nearly $36 billion a year.[4] Canadians for Tax Fairness have identified a further $62 billion a year that could come from making it harder to use tax havens, implementing inheritance taxes, and increasing corporate and investment income taxes.[5] Groups like Resource Movement, made up of over two hundred wealthy individuals in Canada, are calling for a much higher wealth tax,[6] and still others, like British environmental activist and writer George Monbiot, are calling for a ceiling on maximum wealth, after which everything is expropriated.

No matter which proposal we follow, it's worth noting that the means through which the superrich and corporations accrue their wealth are generally not compatible with a just and decolonial world, especially one that will avoid the worst of climate change. While initial corrective taxation measures are needed to redistribute wealth, ultimately we need to change how and for whom wealth is produced in the first place. The overall goal is a transition to a democratic economic system that operates within the bounds of nature and distributes resources more equitably. In other words, if we are successful, in the long term we will build a society where we collectively produce the goods and services we need, cutting corporations and the superrich out of the economy where possible. If we don't allow corporations and the superrich to accrue excess wealth in the first place, there will be no need to extract it back.

### Green Grids
Canada's energy transition must prioritize a shift to 100 percent renew-able electricity and produce enough of it to support the growth in demand as transportation, home heating, and many industrial processes become electrified. As we established in chapter three, the climate emergency timeline precludes using fossil gas as a bridge fuel. If we are going to limit global heating to 1.5°C, we must move aggressively towards 100 percent renewable sources. This is no small task. Renewable energy is less dense than fossil fuels, so its production can potentially require

a bigger spatial footprint. But renewable infrastructure can also often coexist more easily than fossil fuels with other land uses—for example, solar panels can be mounted atop buildings and parking lots, and renewable energies don't contaminate soil through spills and leaks.

In order to ensure this renewable transition is just, we must keep our electrical utilities in, or return them to, public hands, a central demand of many climate organizers associated with the Green New Deal in the United States. Public electrical utilities could reinvent themselves as truly public services, democratically run to prioritize aggressive renewable transitions and to provide energy security to all residents. Rather than serving private interests or acting as sources of government revenue, utilities could provide services including energy efficiency retrofits (reducing the overall need for electricity) and small, distributed, renewable installations that would allow for the production of renewable power in local neighbourhoods and remote locations. Our publicly owned utilities, properly resourced, could act as key infrastructure for heading off the climate emergency before us. By prioritizing long-term investment in renewables as part of a broader future vision, rather than focusing on cost recovery or turning a profit, these utilities could focus on enhancing equity and affordability and moving to a system of dual governance with Indigenous Nations and peoples. Community self-generation could be enabled by and integrated with provincial or regional grids, making both more resilient. The exact electrical mix would be a product of what makes sense locally and regionally, but a focus on funding energy efficiency would first help to reduce the electricity needs in many communities.

Transitioning to renewable energy also means restructuring electrical grids to become much more decentralized. Energy democracy advocates have noted this transition provides a good opportunity to win more distributed democratic control of the energy system.[7] Fossil fuels require large, centralized infrastructure and power plants as well as a top-down electrical grid structure. This makes them suited to ownership by large corporations or state-owned enterprises. In contrast, renewable energy grids are most affordable and resilient when they prioritize the local provision of distributed, relatively small-scale, renewable energy. This means redesigning grids so that many dispersed renewable projects are interconnected with other parts of the grid, where energy flows in both directions, and where energy storage and connections between regional grids ensure consistent access everywhere. It does not mean building renewables only at the scale of rooftop solar installations—the largest solar plant in Canada is 100 megawatts, providing energy to about fifteen thousand households on an average day, a scale still much smaller than most fossil gas power plants in the country. While this redesign is underway, there is an excellent opportunity to reimagine governance at the same time and give communities more direct democratic control and ownership of the energy system.

Solar power is especially suited to being installed, owned, and governed at small scales. Imagine solar cooperatives, democratically run, allowing communities to save through collective procurement and operating on a variety of scales. Other sources of renewable energies will require larger investments, but compared to fossil fuels, they can still be more distributed and democratically governed. Geothermal could operate on the level of neighbourhoods, and small wind farms could supply rural communities. The often high-carbon and Indigenous rights–impeding impacts of hydroelectric developments are drastically reduced when built at a small scale. There has been a surge of run-of-the river hydro projects (that don't require the significant landscape alteration and flooding associated with larger projects) in British Columbia, some of which are owned by First Nations. In general, smaller and more dispersed energy generation will have less impact on communities fighting environmental racism, because there will be fewer big polluting projects that can be concentrated in poor and racialized communities.

However, democratizing our energy sources and governance should not be restricted to self-sufficiency in local communities and neighbourhoods. More centralized and publicly owned green energy infrastructure will be needed as well—in particular, more regional electrical grid connections to help communities sustain each other through interdependence, as well as large-scale storage projects. It is crucial, then, that dual jurisdiction, good, unionized jobs, and bottom-up processes that prioritize social justice and Indigenous rights are also applied to these larger infrastructure projects that a green energy transition calls for.

There is currently a mix of ownership of electrical utilities across Canada, but even the public monopolies are not operating in ways that are democratic or take the public interest to heart. These public monopolies often fail to prioritize clean and low-cost power for residents. As just one example, Saskatchewan's Crown electrical utility, SaskPower, is still largely reliant on coal and fossil gas for its energy mix, yet in 2019, the utility and the Ministry of Environment decided to discontinue their net-metering and solar rebate programs, which were helping to incentivize rooftop solar installations. Claiming that these programs were too costly and were being cross-subsidized by other ratepayers, SaskPower has effectively arrested the growth of distributed solar to the grid. While there may be some merit to its claim that only wealthier residents could afford the solar installations, the real story is that distributed power challenged SaskPower's business model, which relies on centralized production and distribution from large (and fossil fuelled) sources to customers who buy the power. Yet, as a publicly owned utility, there's no reason SaskPower couldn't be mandated to prioritize affordable green energy and support local communities in generating their own power. As the international Trade Unions for Energy Democracy coalition has also cautioned, if we don't prioritize the public good and ensure democratization of our public utilities, interests supporting renewable distributed energy are likely to push, instead, for the privatization of utilities, which would hurt workers and raise prices for low-income communities.[8]

To build out green energy grids, we must avoid the impacts of large "green" megaprojects exemplified by hydro development in Manitoba, British Columbia, Quebec, Labrador, and elsewhere, which have displaced Indigenous communities and destroyed aquatic and terrestrial habitats that supported hunting, fishing, and trapping economies. In stunning reporting on the historical and ongoing impacts of hydro development in Manitoba, APTN highlighted the story of Gerald McKay of Misipawistik Cree Nation, whose community has suffered long-term impacts associated with a large dam at Grand Rapids. In 1969, when the dam began operating, McKay's father lost his sources of livelihood. McKay told APTN:

> In 1969, I think there was mercury in the fish and hydro denied that it was them and then they shut fishing down and there was no compensation for anybody. We all depended on my dad's income to eat, and we couldn't eat the fish anymore. That's just the way it was and so my generation would remember all that stuff, but there's kids growing up now that have no idea what was here before.[9]

Many Indigenous Nations and communities in hydro-affected areas have still not been compensated for the loss of lands and livelihoods from large hydro projects. When they are consulted in advance of new projects, the consultation does not live up to the principles of free, prior, and informed consent, as laid out in UNDRIP.* The Muskrat Falls project in Labrador, which came online in 2020, is a recent example of Canadian governments failing to recognize the rights of Indigenous peoples to consent, or not, before a project proceeds. The Labrador Inuit have serious concerns over the impacts of the 824-megawatt dam, including methylmercury contamination, which the provincial government has refused to address.[10]

It is clear, then, that there can be no new large-scale green energy developments unless they are the product of a genuine process of free, prior, and informed consent. This would require first that Canada give up its claims to jurisdiction and recognize Indigenous authority over the full extent of the traditional territories of each Indigenous Nation or people. Canada could then enter into dual governance arrangements over lands and resources where they are of mutual concern. For example, as we described last chapter, instead of settler governments imposing their will without involving local communities, regional boards and committees could be set up to help direct public utilities. These bodies could be made up not of wealthy elites but of rank-and-file workers and local community residents who would work hand in hand with Indigenous Nations, respecting their jurisdiction. Greening the grid while righting relations will mean redistributing authority and governance.

## Housing and Residential Energy Use
Implementing a just transition will require us not just to change how we produce and distribute energy, it will also involve rethinking how we consume energy and how we structure the places where we live, whether they are urban, rural, or remote communities. In 2019 buildings contributed 12.4 percent of Canada's GHG emissions.[11] Remaking our homes so

---

*   Even in the case of the James Bay and Northern Quebec agreement of 1975, a modern treaty between the James Bay Cree and the Northern Quebec Inuit, the Province of Quebec, and the Government of Canada that allowed for hydro development in exchange for land, hunting rights, financial compensation, and other provisions, it can be easily argued that the Cree and Inuit were not free to say no. In fact, the road and dam construction had already begun before the negotiations, and the communities felt the development was inevitable. In this respect, big green megaprojects can also trample on the rights of Indigenous peoples.

that we can consume less energy is essential to energy transition, and it is also a perfect opportunity to improve the state of housing in so-called Canada and ensure that everyone has a safe place to call home. As we begin to retrofit and build out new social housing, we'll also have to adapt to a changing climate. Investments in climate-resilient green infrastructure can even out the burdens of climate impacts and ensure that everyone has a safe place to call home. Investment should be made first in communities that have faced marginalization and are most susceptible to climate change impacts. We will also need to change how and what we build to ensure new infrastructure is climate resilient—for example, we need to build living shorelines with the capacity to mitigate flooding, rather than sea walls; urban greenspace and buildings that reduce the heat island effect; and infrastructure that is better capable of managing heavy rainfall and periods of intense heat and drought.

Applying our first principle to transition, we imagine a near future where Canada recognizes Indigenous sovereignty over all matters, including housing, and transfers adequate funding to resolve the Indigenous housing crisis on reserves and in cities. This is not a matter of a handout. As the Yellowhead Institute's *Cash Back* report makes clear, Canada is bankrolled by Indigenous lands, resources, and Nations, and it's time to compensate Indigenous peoples for the theft, as partial reparations.

New funding could be invested into housing in Indigenous communities in ways that decrease living costs for residents by instituting deep energy retrofits for efficiency and ensuring that new housing, which is desperately needed to alleviate overcrowding, is constructed using local materials and built to the standards of near-zero. Near-zero energy building design includes increased insulation, an airtight building envelope, efficient mechanical systems, and on-site renewable energy generation.[12] As exemplified in APTN's Power to the People series hosted by Melina Laboucan-Massimo, Indigenous Nations are building distributed and utility-scale renewable energy projects to get off diesel power, increase the comfort of homes, and reduce energy costs in their communities. For example, between 2015 and 2016, the Haida village of Skidegate (population 900) in Haida Gwaii installed 360 heat pumps in band members' homes and 150 kilowatts of solar production on community buildings.[13] The heat pumps have increased comfort levels and reduced energy costs for home heating and cooling. These projects also create short- and long-term jobs, though it is also true that jobs in the mainte-·

nance and repair of diesel infrastructure are lost in the process. Of course these projects are currently taking place within the confines of Canada's laws and policies. Once Indigenous jurisdiction is properly recognized, such projects could do even more to support Indigenous economies and allow for practices of self-determination.

Putting energy-efficient social housing at the centre of green infrastructure planning will provide tangible benefits to people's lives. This contrasts with the federal government's 2030 Emissions Reduction Plan, which incentivizes individual home owners who can afford it to voluntarily retrofit their homes with the help of significant public funding through government rebates and grants.[14] Instead, an approach that centres social housing can begin to undo many decades of broken housing policy and provide special attention to reworking jurisdiction and funding with Indigenous peoples. Over the last decades, real estate markets have become heavily speculative, with investors now making up the largest portion of home buyers in Ontario, and more than one-fifth of multi-family rental properties being held as financialized assets called Real Estate Investment Trusts (REITs).[15] Meanwhile, the stock of social housing has diminished in many jurisdictions after significant offloading of responsibility from federal and provincial governments. This has caused home prices to nearly quadruple in the last two decades, giving landlords cover to hike rents as well.[16]

The situation on reserves is even worse; the federal government has failed to deliver adequate housing, resulting in overcrowding, serious disrepair, and unsafe living conditions. It is worth pointing out that Indigenous peoples were not allowed to purchase their own homes on reserve until 1995, and that, according to Wetchie, First Nations are only funded at $80,000 per house to build on reserve. Obviously, $80,000 does not reflect the true costs of building, especially in remote locations, and keeps housing on reserves substandard. Moreover, many reserves still suffer from a lack of clean drinking water in their homes, despite the federal government's repeated promises to eliminate long-term drinking water advisories—in February 2022, thirty-six long-term drinking water advisories were in effect in twenty-nine communities.[17]

Recognizing Indigenous jurisdiction in housing and transferring the necessary resources to fund it on and off reserve is the first step for Canada. In places like Vancouver and Burnaby, First Nations are already buying up land through the right of first refusal when provincial or federal governments dispose of it—and they are partnering with developers to build market and social housing.[18] But First Nations should not have to buy back land that was stolen from them in the first place by Canada, and this should not become an opportunity to line the pockets of private developers who are already profiting enormously from Canada's housing bubble, government incentives, and permissive municipal zoning and regulation.

In addition to recognizing Indigenous jurisdiction over housing, there is a need to transfer land and cash back to Indigenous peoples so that injustices such as poor housing can be addressed. This can be accomplished even in urban areas where private ownership of land dominates. In the Bay Area of California, the Sogorea Te' Land Trust is securing a land base for the Chochenyo and Karkin Ohlone, whose territories were taken in favour of urban development.[19] The Indigenous, women-led land trust accepts donations of cash and land from individuals and organizations, transferring title back to the Chochenyo and Karkin Ohlone peoples. The Sogorea Te' lands are used for multiple purposes, including food sovereignty projects that grow food and medicines, and ceremonial and cultural practices. But there's no reason they couldn't also be used for housing projects.

Beyond the wealth and corporate taxes discussed above, the Yellowhead Institute, among others, has pointed to the need for a strategy for funding infrastructure deficits (like housing and water) on reserves, as well as language programs and education funding.[20] The Yellowhead Institute suggests that Indigenous peoples could draw up land leases and serve them to cities, provinces, and the federal government, since "'Crown Lands' are based on the legal fiction of the Crown's underlying title to all lands in Canada."[21] Indeed, both the recognition of Indigenous jurisdiction over housing and the transfer of adequate resources and land base to redress the theft inherent in ongoing colonization are required to realize safe, affordable housing that is fit for the climate emergencies upon and ahead of us. There is no scarcity of resources to rectify the injustices.

But housing insecurity is not only produced by colonialism; it is also the product of leaving the allocation of a basic right up to markets and allowing landlords and other private interests to profit from housing. Our second principle of transition supports the idea of removing housing, as much as possible, from the market. This idea is gaining steam as part of proposals for a Green New Deal in the United States. Imagine safe, secure, and energy-efficient housing, recognized as a human right and funded as a social good. This is what the authors of *A Planet to Win* call for in the United States—a homes guarantee that could play a central role in decarbonization, job creation, and racial and economic justice.[22] High-quality public housing, built by unionized tradespeople under direct contract from governments, would serve a mix of income groups and people of all races, avoiding the stigmatization and marginalization of urban housing complexes that marked previous rounds of public housing projects. Inspiration could come from places like Vienna, where still today almost a third of housing is city-owned and another third is cooperatively owned. As described in *A Planet to Win*, Vienna's social housing is beautifully designed, creates hubs for cultural and other public services, and fosters a working-class residential politics that is in stark contrast to conservative currents in other areas of the city and country.

Current policies incentivizing developers to provide below-market rental units are not working; they are subsidies to developers and, given the high market rental rates across the country, do not provide housing that is truly affordable. According to Statistics Canada, more than 235,000 people in Canada experience homelessness in any given year,[23] a figure that is likely an underestimate because of a lack of visibility—unhoused people may be couch-surfing at the homes of friends and relatives or living in inadequate conditions. And among those who are housed, the Canadian Mortgage and Housing Corporation found that approximately 1.7 million Canadian households, or 12.7 percent, were in need of housing assistance because their before-tax income was not sufficient to access acceptable local housing,[24] meaning their housing costs over 30 percent of their before-tax income. Roughly 20 percent of Canadians in 2016 were living in states of energy poverty, meaning they struggled to heat and cool their homes and power their lights and appliances, spending more than 6 percent of their after-tax household income on home energy services.[25] The COVID-19 pandemic has only worsened this picture, and accelerating the climate crisis will plunge more people into energy insecurity and render more housing stock substandard.

There are already growing efforts to address these problems, in large part because COVID-19 pushed a housing system that was already buckling over the brink. Growing houseless populations in cities across

Canada built new encampments to escape unsafe and underfunded shelter systems. This crisis prompted stronger community networks for mutual aid—including the Encampment Support Network in Toronto and Camp Pekiwewin in Edmonton (amiskwaciwâskahikan)—that have agitated powerfully for many of the solutions we talk about in this chapter and helped protect encampment residents from violent, multi-million-dollar police evictions.

Now it is time to refocus our efforts on eliminating the housing crisis in Indigenous communities and on a Canada-wide homes guarantee. While governments have never prioritized good housing in Indigenous communities, there is a history of social housing provision that is worth learning from. Some sixty years ago social housing was a significant priority for the federal government, with many provinces also playing a crucial role. Amendments to the National Housing Act in 1964 and the creation of the Ontario Housing Corporation in the same year, for example, led to a significant increase in social housing in Ontario, with social housing production increasing ten-fold. This was on top of an earlier postwar period of modest growth that resulted in a public housing stock of six thousand units in Toronto.[26] Indeed, Prime Minister Lester B. Pearson's government saw social housing as key to urban renewal, and not just in Toronto or Ontario. According to Gregory Suttor's University of Toronto PhD dissertation, during the 1960s public housing was not stigmatized and was built to house people of a variety of socio-economic status as part of a policy of managed urban growth. "Ideas on mixed-density, nodal suburban development were not focused, as they are today, on urbanity of urban form, environmental objectives, or transit-supportive density; they were explicitly about a mix of prices, tenure, social classes, and stages of the life cycle."[27] We now need to reconfigure the built form that we inherited from this era (which extends well beyond social housing) to ensure socially just and decarbonized cities. But it is worth noting that houselessness at the scale we know it today did not even exist until many of these housing programs were wound down in the 1980s.[28] We could return to building out social housing today to more or less eradicate the problem of people being unhoused.

The social housing that still exists in Canada (mostly built in the era just described) is in dire need of repair and deep energy retrofits, and since governments have largely divested themselves from social housing provision, there is a desperate need to build more social housing in rural, urban, and remote areas. Today, instead of incentivizing private developers to build affordable housing without requiring energy efficiency, the Canadian government could invest directly in building the social housing infrastructure of the future. Using legislation similar to what Democrats Bernie Sanders and Alexandria Ocasio-Cortez recently introduced in the United States under the banner of a Green New Deal for Public Housing, the Canadian state could hire unionized tradespeople directly to retrofit and build out new social and public housing across Canada, creating a large number of good jobs that respect the integrity of the trades. This housing should be built to the highest standards of energy efficiency, which will reduce overall energy use as well as residents' energy bills. It also needs to be housing that is resilient to the impacts of climate change. Bulk procurement for construction will see large economies of scale and support good jobs, and this publicly owned housing would ensure protection for common people as infrastructure becomes more vulnerable to the effects of climate change and, consequently, insurance rates for housing and other infrastructure rise.

Other corrective policies, borrowed from other countries, should be introduced to constrain private profit from housing. Venezuela, for example, has delivered over three million new dwellings to its residents through its Great Housing Mission, a community-led social housing program.[29] A tax on vacant and second homes could be reinvested in public housing, reducing the number of underused homes in private hands, and a cap could be placed on the number of units a landlord can own, as is the case in Berlin. But if Canada wishes to achieve significant levels of public ownership, housing assets owned by large corporations will also need to be removed from private ownership through expropriation.

Building high-quality public housing through government contracts and unionized labour for a variety of income groups and people of all races will help to gain momentum for the large-scale changes that are needed to address the climate crisis. Most importantly, recognizing Indigenous jurisdiction over housing and properly funding it, as well as providing a homes guarantee supported by abundant, beautiful, and publicly procured public housing, will provide real tangible benefits that will change people's lives. These are the types of policies, investments, and recognition of Indigenous rights that have the potential to rally wide swaths of the public behind energy transition.

## Clean Transportation for All

Our vision of the good life includes safe, affordable, green transportation for all, owned publicly, and governed in ways that recognize Indigenous jurisdiction and sovereignty. For people living in rural and remote locations, imagine affordable intercity transit using buses, trains, and smaller vehicles, that connects all communities across so-called Canada in an integrated transit network fuelled by renewable energy and governed jointly with Indigenous peoples. Picture seamless and safe routes that pass through regional hubs and connect to major centres so that people can visit loved ones and access a diversity of services and experiences. Within cities, imagine getting around town quickly and without cost on low-carbon transit systems or walking, wheeling, or cycling on safe, accessible sidewalks and corridors that connect people to the services they need. Longer term, imagine our built environments being reworked away from the private automobile and ever-sprawling suburbs to instead prioritize accessible and affordable mass transit. Building out public transit systems through shared governance will produce thousands of good, unionized jobs and change how we relate and connect to one another.

We need to prioritize green mass transit because fossil fuelled transportation is currently a large contributor to Canada's greenhouse gas emissions, 25 percent of national emissions in 2019.[30] Between 1990 and 2019, GHG emissions from the transport sector grew by 54 percent.[31] Much of this increase was the result of more freight truck traffic as well as the growth in passenger light trucks.

Mainstream policy approaches to the problem of transportation emissions have so far been focused on incentivizing electric vehicles (EVs) and subsidizing EV charging stations. This is the strategy taken in the federal 2030 Emissions Reduction Plan, which says very little about public transportation (though it does announce continued funding), while focusing on incentives for light and heavy duty ZEVs (a category of "zero emissions vehicles" broader than EVs, including plug-in hybrids and hydrogen fuel cells) and investing in charging infrastructure through the public-private partnership model inherent in Canada's Infrastructure Bank.

EVs and other "zero-emissions vehicles," despite their marketing, are not without emissions. Much energy is required to manufacture these vehicles, including for steel and batteries. In fact, over the life cycle of an electric vehicle, it may produce around 30 to 70 percent of the emissions of a combustion-engine vehicle, depending on energy sources for manufacturing and charging.[32] And this myopic attention to electrification of private automobiles has significant social implications—EVs cost more than gasoline or diesel powered vehicles, while many people cannot afford to own and maintain a vehicle of any kind. Furthermore, if green transportation policies focus primarily on electrifying private automobiles, they will fail to deal with the worrisome growth in energy demand, as research shows that people are transitioning from smaller, more efficient cars to larger SUVs, vans, and trucks. In other words, incentivizing individuals to choose EVs benefits only a certain class of consumers and fails to provide a public option for all. It also fails to make the infrastructural and urban-level changes that would improve the lives of so many. Ultimately, we need to decrease the energy demanded by private automobile users as a matter of energy conservation, but also as a matter of global justice. In many parts of world, especially in the Global South, environments and communities are torn up in the frenzy to secure and extract lithium for batteries and other metals to replace the more than one billion combustion engines currently driving the world's roads and highways. Mass transportation, not the private automobile, must be at the centre of visions for a better world.

Applying our two principles, we must first consider how Indigenous rights and jurisdiction could be recognized, and Canada's jurisdiction scaled back, in order to provide clean transportation. Connecting all communities across Canada in an integrated, fossil fuel free transit network through genuinely shared governance cannot be accomplished through private markets, since shared governance will require planning and coordination with the explicit purpose of connecting remote communities and ensuring safe transportation for all. Clearly Indigenous communities have a lot to gain from improved and safe public and collective transit within cities, between cities, and to remote communities. British Columbia's Highway of Tears, where dozens of Indigenous women have gone missing over the last fifty years, with many more experiencing violence along the 725-kilometre corridor of Highway 16 between Prince George and Prince Rupert, is only one example of the desperate need to provide safe intercity transit.[33] A national intercity and rural bus and rail network will have to be attentive to each Nation's priorities and wishes and negotiated through genuine processes of free, prior, and informed consent.

Unfortunately the state of intercity and rural transit has only deteriorated over the last decades as both public bus services (such as the Saskatchewan Transportation Company) and private ones (such as Greyhound) have closed down operations in Canada. Passenger rail service has also declined precipitously; cuts were imposed by federal governments in 1981, 1990, and 2012, and the privatization of CN in 1995 hurt service as well.[34] This has left the state of intercity transit highly uneven across the country, with investments made in major corridors like Montreal-Toronto and other routes left unserved. Passenger rail service in the Prairies was better in the 1920s than it is in the early 2020s. Many intercity routes within Canada are simply unprofitable for private operators, so clean transportation for all will need significant public investment. In order to put justice at the centre of a clean transportation strategy, author James Wilt proposed in his book *Do Androids Dream of Electric Cars?* a future where "vehicles and systems are publicly owned, driven by professional and preferably unionized workers, and operating for socially beneficial purposes—not private profits."[35] A national intercity and rural bus and rail network will need significant public investment to ensure good and equitable service across the board and prioritize the needs and jurisdiction of Indigenous peoples. Decisions about high-speed routes, train versus bus routes, and the treatment of freight versus passenger service will all need to be worked out across jurisdictions.

Within cities, public transportation should also be a priority for energy transition. While cities often prioritize the electrification of transit fleets in their green energy transition strategies, the approach needs to be much wider. Ultimately, we need cities that are built to facilitate, prioritize, and make more accessible modes of transportation that are not fossil fuelled (transit, cycling, scooting, walking, and so on). As many researchers have documented, the sprawl that is characteristic of low urban density and ever-expanding suburbanization is facilitated by the private automobile and perpetuates an urban form and culture entirely dependent on automobility. While public transit ridership has generally been decreasing across North America (with a few notable exceptions such as Vancouver[36]), a green energy transition needs to reverse this trend, since collective transportation is much more energy efficient than private electric vehicles.

Recall here that there are justice implications with ever-expanding green energy consumption, so that reduced energy consumption must be part of a green transition. Moreover, as a matter of social justice, it is important to shift investments from costly car infrastructure to the infrastructure that marginalized communities can most benefit from. Calls for universal fare-free transit are gaining steam across North America, and could make a real difference in many people's lives. Through focus groups about Regina's renewable transition, one of this book's co-authors, Emily Eaton, alongside Simon Enoch, found a great deal of enthusiasm for fare-free transit. Despite Regina being considered a car-dependent city with a frigid climate, participants stressed that a large number of youths, seniors, people living in poverty, disabled people, and newcomers do not have access to cars—they take public transit only when they can afford it and otherwise walk or bike long distances, or have to stay home.[37] Bringing in fare-free transit in the city would produce an instant benefit for these groups and increase ridership, which will perpetuate a culture of public transportation use, increase safety for transit users,. and help make the case for more routes and better frequency.

Ultimately, the structure of cities needs to change if more people are going to get on public transit and take fewer trips in their private automobiles, even if those vehicles are electric. Fortunately, the field of urban planning has already laid out solutions, which include increasing the density of urban areas; ending suburban expansion; siting amenities such as food, recreation, and shopping within residential areas; reducing the availability of parking; designing streets and sidewalks with pedestrians, cyclists, and urban transit in mind; increasing the frequency and accessibility of public transit; initiating rapid transit routes; improving cycling and pedestrian infrastructure; and much more.[38] The solutions are readily available and backed by a significant amount of research, but municipal governments have succumbed to the power and influence of industries (such as real estate, developers, and fossil fuels) that want desperately to continue expanding suburbs and private automobiles in order to increase dependence on their products.

Winning good and free transit for all will thus require significant mobilization of communities, but this struggle has the potential to unite many different advocacy groups including anti-poverty, disability, Indigenous rights, transit riders, and labour unions. And the work has already started. In Edmonton, Free Transit Edmonton launched in 2019 as a working group of Climate Justice Edmonton that our co-author Bronwen Tucker organizes with. Working with the Edmonton local of the Amalgamated Transit Union, disability justice organization Self Advocacy Federation, Black Lives Matter YEG, encampment support groups, and others towards the elimination of fares altogether, it had in the first three years of its existence already won a 40 percent decrease in fines for fare evasion and prevented a proposed fare increase.

Similar free transit campaigns and transit riders' groups are emerging across Canada, both in large cities (like Toronto) and small ones (like Saskatoon), and they are allying with transit operators and maintenance unions, anti-poverty groups, and others to improve transit and maintain good, unionized jobs. In 2019, for example, Toronto Transit Commission electrical workers, organized through CUPE Local 2, called for free public transit and a campaign of mass strikes and protests to bring down Premier Doug Ford, a premier who has attacked public services and unionized workers and cut funding to municipalities.[39] In their news release, they called transit fares a tax on the working class and the poor. In the same year, Amalgamated Transit Union Canada initiated a campaign calling for a "national public intercity transit service as part of a Green New Deal for social, economic and environmental justice and tangible reconciliation with Indigenous peoples."[40]

* * *

Green energy infrastructure; safe, secure, and social housing; and clean transportation for all are tangible goods that we can deliver collectively through well-paying unionized jobs while righting the relationships that allow for the coexistence of Indigenous and settler peoples in these lands. Building out this infrastructure while scaling back Canada's jurisdiction and recognizing Indigenous rights provides a hopeful and productive way to fight climate change while delivering benefits that will make many people's lives better. Mobilizing the masses around this agenda will be much easier than getting people to love carbon taxes or embrace incentives for industries that have already caused a lot of damage. In fact, this vision takes its lead from grassroots initiatives, movements for Indigenous rights, and labour organizing. It is an agenda for people and the planet, rather than for green capital and the political elite.

# 5.
# MIYO-WÎCIHTOWIN

## Uniting to Build a Caring Economy for All

On May 3, 2016, "Reyna," a live-in caregiver in Fort McMurray, Alberta, opened the door to a dark, orange sky with ash falling like snow. Realizing that a massive wildfire was approaching, she gathered food and clothing for her employers' young daughter and evacuated the city with the girl and her parents. She spent the next ten days with the family helping them get through the evacuation process until they were allowed to return to Fort McMurray, and the weeks after it helping repair their home and possessions on top of her regular duties.

The 2016 wildfire that swept through Fort McMurray and forced the evacuation of 88,000 people is well known, but the care workers like Reyna who were on the front lines of it are much less talked about. We heard Reyna's story through Emma Jackson, an organizer with Migrante Alberta and Climate Justice Edmonton who, while researching the wildfire, met with Reyna and many other Filipina migrant caregivers.[1] Most of the caregivers Emma spoke with were in Canada under the federal government's Caregiver Program. Under normal conditions, this program already makes migrant workers incredibly vulnerable by tying their status in Canada to a single contract in a way that means refusing unfair conditions would risk their deportation. But as the economic implications of the fire hit the families who employed them, conditions for many of these workers got much worse. In Reyna's case, her employers demanded she pay thousands of dollars to help them cover the cost of evacuation, cut her hours severely (from forty-four to twenty hours per week) despite contract terms guaranteeing her thirty hours per week, and were controlling about any other work she took on to help her cover costs for herself and her family in the Philippines.[*] After providing similar emergency care and facing some of the wildfire's most direct dangers, in its wake, many other migrant caregivers also faced dramatic pay cuts, outright dismissal, and deportation.

We don't need to look far to find other examples of how Canada's current economy undervalues care work. "Care work" is simply the work of caring for others, whether paid or unpaid, formal or informal. It is the work—child-rearing, cooking, cleaning, emotional support, eldercare, and much more—that makes all other work possible. But rather than being highly valued as life-supporting, care work in Canada is almost always underpaid or not paid at all, and it is disproportionately carried out by women and people of colour. Great swaths of our childcare, eldercare, and healthcare systems are privatized and run as for-profit ventures, creating jobs with terrible working conditions that are left to the least powerful and least resourced members of our society. Unpaid and informal care work is also increasingly squeezed as more people juggle multiple jobs and rising costs of living that limit the time we have to support each other.

And these inequalities of care are often made much worse when we enter times of crisis. The COVID-19 pandemic reproduced many of the same patterns as the Fort McMurray wildfire. Many care workers were doing essential work caring for children, the elderly, and immuno-compromised people, all while facing some of the greatest COVID-19 risks and being afforded little protection. Care work is also very gendered and racialized, as were the impacts of the pandemic. The Canadian Centre for Policy Alternatives found that women "have been stepping up to shoulder a huge pandemic increase in unpaid labour and caregiving and stepping back from paid employment."[2] Low-wage workers, overwhelmingly women, highly racialized, and facing the greatest barriers to employment, suffered the largest share of job losses and reduced work hours.

Let us start this chapter with a reminder that this situation is far from inevitable. The Indigenous economies, laws, and social systems of care we will discuss show us that we can build a society with care at the centre

---

[*] To complicate things further, during the evacuation and shortly after, Reyna was forced to go on employment insurance to help alleviate her employer's expenses. They expected her to collect EI while doing care work for them, so they did not have to pay her salary, and expected her to pay the expenses of the evacuation from her EI payments.

of it. In fact, building such a society is a critical part of a just transition. A key first step is restoring Indigenous sovereignty and economies so that Indigenous Nations can rebuild these systems on their own terms. Centring care also means building universal public services—from eldercare to mental health care to daycare—to ensure that care is abundant and available to all and that the life-supporting jobs in these sectors are decent, high-quality jobs. More broadly, it also means ensuring decent work and basic needs are available to all through policies like livable minimum wages and full immigration status for all. Winning these policies will not only ease the uneven burden of unpaid care work, but they are also essential elements of climate action. By creating good, low-carbon jobs, an abundance of care, and meeting everyone's basic needs, we will lower emissions and become much more ready to respond to the floods, fires, heat waves, and other climate-driven crises we can expect in years to come.

Here, we draw on nehiyaw (Cree) understandings of economies of care, as co-author, sociologist of family and work, and former union researcher Angele Alook is nehiyaw. This is one Indigenous perspective; if this chapter were written from a Mohawk perspective, a Dene perspective, a Métis perspective, or an Inuit perspective, it would be just as insightful and valuable. Settler readers are encouraged to be good allies to the Indigenous peoples in their home territories and learn more about economies of care based on local Indigenous relationships with the land.

## The Capitalist Colonial Economy of Canada
Ojibwe (Anishinaabe) activist and author Winona LaDuke and settler geography professor Deborah Cowen call the Canadian economy a Wiindigo economy.[3] The Wiindigo is a cannibal creature in nehiyaw and Anishinaabe legends;[†] stories of him are told to teach the young ones lessons. In these stories Wiindigo consumes his family and his community, and he is only finally stopped by removing his head from his body. This legend is told in Angele's home community of Bigstone Cree Nation, where the Wiindigo's head, body, and the axe used to kill him

---

† It should be noted that according to Cree spiritual practices and Cree laws, one should be very careful in what words are spoken out loud; what you say has an impact on the living world around you. Therefore, even saying the name of the wîhtikow (Cree spelling of the word Wiindigo) out loud could bring forth demise. This demonstrates the power of using this evil figure in this book. There is even a special protocol for using this story.

are all buried in secret separate locations. Given how evil this murderous cannibal was to the village, the location of his death and his body parts is withheld, lest Wiindigo return to do more harm. In Anishinaabe legend this figure "destroys itself through addictive indulgence in its craven desire."[4] In Angele's family this was a story told as a warning against being greedy and selfish, to make children aware of evil spirits, and to instill a fear of the evil that comes from greed and gluttony.

It is easy to relate the Wiindigo figure to Canada's economy. As we have already seen, the history of Canada is one of corporations accumulating wealth through the theft of resources from Indigenous lands and peoples both on Turtle Island (North America) and beyond. In fact, before Canada was a country, much of it was a corporation, a gift from the English Crown to the Hudson's Bay Company in 1670. These lands were then bought by Canada in 1869 for $60 million dollars (in today's valuation). The traditional territories of nehiyaw, Siksikaitsitapi (Blackfoot Confederacy), Nakoda Oyadebi (Assiniboine), Dene, and other Nations were exchanged on paper between colonial parties but were never ceded by any Indigenous Nations.

Throughout Canada's history, Indigenous economies have been pushed aside and criminalized in favour of settler agriculture and the extraction of resources, building out a web of infrastructure, sacrifice zones, and urban settlement that has undermined traditional Indigenous economies of hunting, harvesting, and agriculture. Settlers were allowed to over-hunt the bison to near extinction, decimating a major part of Indigenous prairie economies and a chief source of protein. This was written off as merely "collateral damage of colonization."[5] With their own economies consumed by capitalist colonialism, when Indigenous peoples have attempted to participate in settler economies, they have faced systematic exclusion and racist policies like the Indian Act's ban on the "sale of agricultural products by 'Indians,'" which was implemented when nehiyaw farmers successfully competed with settler farmers in the early 1900s.[6] This destruction extends past Canada's borders, with Canadian extraction companies ruining ecosystems and local livelihoods. From the Philippines to Guatemala to Ecuador, Canadian destruction and militarization has forced local populations to migrate, ironically oftentimes to Canada, where they are then forced into precarious, underpaid work to subsidize this broken settler economy.[7]

After destroying Indigenous economies and preventing participation in the settler one, the federal government has maintained control over Indigenous Nations through annual funding agreements that keep Indigenous peoples, especially those in remote communities, in a state of poverty, providing just "enough to keep them alive."[8] This is all done while stigmatizing Indigenous peoples for relying on federal funding. And in response to calls for self-determination, the federal government has offered only devolution of funding, where First Nations are expected to self-administer funds and programming, with less funds than provincially regulated jurisdictions and very little decision-making responsibilities. Throughout this terrible fiscal relationship of colonization, as the Yellowhead Institute's 2021 *Cash Back* report calls it, Indigenous peoples have adapted by participating in mixed economies: the Mohawk built skyscrapers in Manhattan, Algonquin people worked on mink farms, and coastal peoples participated in commercial salmon fishing.

Today the capitalist colonial economy continues to push deeper into Indigenous territories, always in search of new resources to consume, all while destroying forests, water systems, and life along its path, cannibalizing itself—because there is no economy on a dead planet. This hurts everyone in so-called Canada, through environmental destruction but also through a politics of austerity that starves care sectors in order to feed industry with subsidies and deregulation, and a drive for profits that treats workers as disposable. As LaDuke and Cowen write, this Wiindigo economy "denigrates the sacred in all of us."[9]

## Indigenous Laws and Social Systems of Care

To discuss Indigenous understandings of care, we must first look at how Indigenous societies were traditionally structured. Notably, the child's well-being was central to how political, social, and economic systems were designed. Cree/Métis author and respected Elder Maria Campbell explains the traditional social organization of her people in concentric circles, with children in the innermost circle, surrounded by a second circle of elders, then a third circle of women, and finally in the outermost circle, the men of the community. Elaborating on this description, Cree/Métis historian Kim Anderson describes a society that first takes care of children and elders, a system of intergenerational teaching and value in the future generations of the Nation:

In this worldview, children are at the centre of the community. The elders sit next to the children, as it is their job to teach the spiritual, social and cultural lifeways of the nation. The women sit next to the elders and the men sit on the outside. From these points they perform their respective economic and social roles, as protectors and providers of the two most important circles in our community.[10]

In 2012 Angele interviewed Indigenous parents in Edmonton and her home community of Wabasca and saw this model reflected in her conversations. The care and development of children and the mainte-nance of family relations were central to their decisions about school and work.[11] These parents also held strongly that the well-being of their children was closely interrelated to that of their wider family and community. Their children received care from grandparents, and aunts and uncles, who acted as "co-caregivers" with the parents. And parents were able to participate in post-secondary education and steady employ-ment because of this help from co-caregivers. These parents showed they consider their children to be sacred gifts and valued members of the community to be taught and nurtured by grandparents, aunts, and uncles. As Angele's study with these interviews concluded, "There is an understanding that the survival of our ways of being and our future self-determination are contingent on how we care for our children."[12] These contemporary Cree and Métis understandings, which value care work as self-determination, are in line with traditional teachings in many Indigenous societies on Turtle Island and beyond.

If putting children at the centre of social organization sounds obvi-ous or universal, note that in contrast, early Jesuits of New France actu-ally criticized the "excessive love of their offspring" among Algonquian and Iroquoian peoples. Nineteenth-century settler writer Amelia Paget went as far as to note that among the Cree "the love of their children was a particularly pathetic trait in their natures."[13] We can see the reflections of these statements in the precarious state of childcare in Canada today.

In response, Anderson writes that this "excessive love and nurtur-ing" is actually key to the goal of children living a good life, along with freedom to play and learn without punitive, harsh forms of discipline (a cruelty that was introduced in the residential schools, among other colonial systems of oppression). The elders Anderson interviewed about life stages discussed how children were raised by the community, among

many caregivers. Food and love were shared in abundance. The principle of reciprocity in relationships was key: "Youngsters were not simply
passive recipients of care and teaching. They were often helpers to their
grandparents and were given tasks and responsibilities that facilitated
learning."[14]

Speaking to Anderson, Saulteaux Elder Danny Musqua said:

> Every stage of childhood was a celebration because children needed
> to develop a sense of belonging; that sense that "You are important to
> the people." [If] a child doesn't have a sense of belonging and respon-
> sibility as a part of the whole, that's the weakness that will tear up a
> community.[15]

Musqua also teaches that children needed to learn Indigenous family
law and community law, to know the rules and regulations to be a full
contributing member.

Nehiyaw legal scholar Sylvia McAdam (Saysewahum) explains the
law of raising up children in a good way. Below are some of the central
teachings of this law:

> A nêhiyaw child is born with many gifts; nêhiyawêwin (language),
> pimâtisiwin (life), pimâcihowin (livelihood). More specifically, there
> are four gifts given to them: emotional, mental, physical, and spiritual.
> All four have to be in balance with each other, utilizing the nêhiyaw
> laws as the foundation at all times.

> The four gifts are nurtured by the laws, especially the law of miyo-
> ohpikinâwasowin. . . . miyo- means good, ohpikinâwasowin means
> child-rearing or raising. miyo-ohpikinâwasowin directs parents to raise
> their children to become lawful nêhiyaw citizens for their respective
> nêhiyaw nations . . . from cradle to death; their responsibilities do not
> end when the child reaches a certain age. It is a lifelong responsibility.

> Each nêhiyaw child has a birth right that is steeped in the history of the
> land and their kinship with all of creation. They are born into respon-
> sibilities and obligations that will guide them from cradle to death.[16]

McAdam stresses that following the law of miyo-ohpikinâwasowin will lead to decolonization and to a path of self-determination; this will stop the genocide and lead to Indigenous peoples living in a good way. If we can build economies of care based on children living a good life, then this will lead to future generations knowing their obligations to the collective good. One of the key components of climate justice is intergenerational responsibilities and caring for future generations to live on this planet.

Further nehiyaw laws would also contribute to climate justice and a just transition. The Wahkohtowin Law and Governance Lodge bundle teaching explains the nehiyaw law of miyo-wîcihtowin, introduced in chapter one, which means "to get along together [and] having a good relationship with each other." miyo-wîcihtowin "is viewed as strengthening unity while reaffirming the importance of good communication, kindness, [and] respect which may be necessary elements of nation-building."[17]

Another law that speaks to obligation to the collective good is sihtoskâtowin, which is understood as "the act of supporting and helping each other as well as looking out for one another." This law was vital for when Nations "pulled together" as one "to ensure survival and well-being of all community members." For example, community members shared food and resources like firewood, and worked together in harvesting and preparing these resources. According to the Wahkohtowin Law and Governance Lodge bundle teaching, sihtoskâtowin "encompasses collective support for the wellness, safety, caring, and protection of the Nation's members."[18] This law is about taking care of the most vulnerable in the community, often our children and elders, and sometimes those grieving a loss or some other unforeseen tragedy. This nehiyaw law is applicable to building a just transition, since those made most vulnerable to climate threats often have the fewest resources to survive disasters, so we must pull together for them in order to achieve climate justice.

A law that is especially important for climate justice is manâcihtâwin, which "encompasses the principle and related doctrines of respectful relationships between all beings, animate and inanimate." More simply, the straight translation of the word manâcihtâwin is respect.* This law is

---

* According to Ron Lameman, nehiyawewin expert from Beaver Lake Cree Nation, who was consulted by the authors of the book, the straight translation of the word manâcihtâwin is respect. As well, he introduced the term wânaskêwin, which means living in peace and harmony; this is another term to be considered here.

about maintaining a good relationship with the land, defined as "civility, showing respect to all of creation." Cree people understood that respect and reciprocity guided their relationship with the land. A reciprocal relationship with the land means knowing that "if humans treat the land in the right way, in turn the land will provide for humans and uphold their well-being."[19] Alternatively, if we take too much from the land or harm the land, then the land will not be able to uphold humanity's well-being. This is what we are witnessing in the environmental and climate crisis, where we are taking too much from the land for profits, harming the land with pollution, and damaging the planet with fossil fuel emissions. The ecosystems of our forests, mountains, oceans, and atmosphere are out of balance because we have not followed the law of manâcihtâwin. And we have not treated all humans with respect and reciprocity either, for example, large developed nations like Canada consume more goods and extract more fossil fuels than other nations, and yet it is mostly Indigenous and racialized peoples around the world who bear the burden of climate disasters (like the forest fires in Canada's North).

These laws can be used to restore Indigenous economies. Now let's look at how.

## A Just Transition Means Building Care Economies for All

As the authors of *Cash Back* wrote in the wake of the COVID-19 pandemic, "The *Wiindigo* economy—a society built on death—shows why it is settler society that will need First Nation leadership and support to 'build back better.' This time, the new economy must be built on life."[20] The laws of miyo-ohpikinâwasowin (raising children in a good way), miyo-wîcihtowin (good relations and unity), sihtoskâtowin (pulling together for survival), manâcihtâwin (respect and reciprocity with the land), all lead to what Angele refers to as miyo-pimatisiwin in her research—living a good life with balance between family and work.[21] And together, these laws provide a strong basis for the climate-safe, care-centred, and equitable world we want to build going forward.

There are a lot of tangible first steps we can take; here we'll cover four of the most important ones. First, continuing ongoing work to restore Indigenous sovereignty to give space for Indigenous Nations to restore and reimagine their care-centred laws and economies. Second, fighting for policies that will help create good, low-carbon jobs doing care work and make care available to all through universal public services. Some services, like childcare, have been implemented to a limited degree in Canada, while others, such as universal public dental and mental health care, seem further out of reach but would go a long way to improving our well-being and alleviating pain. These are all achievable goals that can be fought for. Third, policies to help restore miyo-pimatisiwin (living a good life with balance between family and work), which can be translated loosely into the realm of the labour movement and specifically the right to "decent work." Per the International Labour Organization, decent work means ensuring "opportunities for work that is productive and delivers a fair income, security in the workplace and social protection for families, better prospects for personal development and social integration, freedom for people to express their concerns, organize and participate in the decisions that affect their lives and equality of opportunity and treatment" for everyone.[22] There are labour rights and other policies we can fight for that will ensure decent work and free up time and resources—especially among the communities made most vulnerable by the Wiindigo economy—to centre care for each other and the land we all rely on. Finally, to enable this work, we must claw back the sectors of our economy most directly blocking these laws from being restored.

### Restoring Indigenous Care Economies

A capitalist colonial economy has caused Indigenous societies to go from economic systems of sharing, caring, respect, and reciprocity with the land, where responsibility to the collective provided well-being from cradle to grave, to a colonial system that displaced Indigenous peoples from their lands and pushed them into economies of poverty and dependency with only enough funding to "keep them alive." Working to restore miyo-pimatisiwin will destroy a capitalist colonial economy that is based on a "death drive" of Indigenous peoples,[23] and restore ethical Indigenous economies that are adapted to our present realities.

Despite Canada's consistent colonial policies, Indigenous economies have persisted and adapted over time to cover a great diversity of activities. Examples provided by the authors of the *Cash Back* report include community-regulated fisheries, feasts and community freezers of wild meat that nourish bellies and hearts, sugar bush camps and salmon harvests, lipstick lines, airlines, and moccasin-making microenterprises. They are also apparent in defund the police movements, harm reduction initiatives, and Friendship Centres. "At their core, what makes them Indigenous economies is that they do not exploit that which they depend upon to live, including people. And they protect a world that is not prepared to value people's time, homelands, and harvests solely in cash."[24]

To further the work of restoring Indigenous economies, we must reclaim Indigenous laws and Indigenous languages, but we also need to return lands to Indigenous communities and recognize the jurisdiction of Indigenous Nations over their territories.[25] Currently, the reserve lands where many Indigenous peoples live make up only 0.2 percent of Canada's land mass. This is too small of a land base to sustain Indigenous economies. We must look beyond the boundaries of the reserve, and beyond the fiscal boundaries of paternalistic funding arrangements that paint Indigenous peoples as unable to manage funds.

Indigenous peoples are model stewards of the land because their economies rely on relations with Mother Earth based on reciprocity and respect. For example, Cree and Blackfoot peoples have traditionally done slow burns of the grass to enrich the diets of big game and maintain the grasslands; now these traditions are being called upon to help manage hotter and faster, and deadly, forest fires. Indigenous peoples can help control the fires that are killing Mother Earth. We are water protectors, land defenders, ocean experts, care workers, wage workers, and excellent at managing resources. Give us land back so we can restore the earth to economies of care and life. In so-called Canada, start by returning Crown lands, which were stolen from Indigenous peoples in the first place.

In the interview excerpt below, former Grand Chief Derek Nepinak talks with Angele Alook and Crystal Lameman about a number of themes: creating hunting management programs where Indigenous and settler hunters can share in the big game, how wage labour is exploitative, how Indigenous peoples combined wage labour and living off the land to survive, and how capitalist colonial understandings of wealth and poverty aren't commensurate with Indigenous understandings of food sovereignty and caring. This discussion illustrates the value of Indigenous ways of knowing, being, and doing in building Indigenous economies of care.

## DISCUSSION WITH DEREK NEPINAK, ANGELE ALOOK, AND CRYSTAL LAMEMAN
### April 9, 2021, Zoom meeting

**Angele:** What has been unjust about livelihoods and access to jobs in the past, and how can we make it just in the future?

**Derek:** When we think about our access to the land and our resources, we also have to balance that with the interests and motivations of settler society that wants to access those same resources for different reasons. For example, I wanted to go to Duck Mountain in western Manitoba and take a moose. We went out with a big group, and it was supposed to be my son's rite of passage, but he didn't find a moose. So I had to take one from the mountain. It's not an infinite amount of moose out on the mountain, there's only a limited number that can be hunted. But we must balance those interests of the rites of passage, those cultural and ceremony spaces that we create, with our relationship with the moose, with other interests. Those other interests include Manitoba provincial conservation interests that want to hit a critical number so that they can start selling moose tags to people to access the resource as well. We've always lived in a space and in a time where we interact with other people that bring other perspectives and other interests into our home territory. So when it comes to livelihood, we have to consider and conceptualize possibly an integrated approach.

In Wisconsin, I saw they developed a cull management regime where the meeting starts with the drum and the pipe ceremony. State officials interface and interact with Indigenous land keepers, and they talk about the state of big game. For example, they talk about the state of fishing in that environment, and they've been able to harmonize and bring together an Indigenous-centred approach to management of the resource. And I've always believed that it could be a viable option if we could find respect. If we could find a way to sit respectfully and to talk about those things that are most important.

Now when it comes to working or jobs, I've always kind of been of the view that we all need resources. We all need ways of finding comfort and a way to live well. First, for a lot of people, that translates into having a job or getting a job and trading their time for money over a fixed period of hours in a day or a fixed period of hours in a week or a month, whatever the case may be. And I've struggled with that because I think that that indoctrination starts early in school for kids. You get your university education and then you get a job. And then you start trading your time for money. I don't necessarily think that that's always been the best way for us.

When we look back on some of the old ways that our ancestors engaged in wage labour, they combined it with ways of living off the land, like hunting. They became the best workers, because they'd work twelve to fourteen hours a day non-stop, and I've seen that still with my uncle. When it comes to hunting, he'll work two weeks in a row, to go out with groups and help skin, help gut, help drag meat out of the bush. That's some of the hardest work that you can ever do. I think there's broader concepts in terms of how we could create well-being for our families, and resources and revenue for our families, that we haven't really spent a lot of time thinking of yet. You know, that's a really difficult concept to deconstruct and then to rebuild, but it could be done.

**Angele:** I often think about this word in Cree when I think about labour: "kitimâkis." It means they are poor. And then I think of the word "atoskewin," meaning to work. And then I think "kitimâkis-owin," meaning being poor or unhappy. To call someone poor in Cree is not the same as in English. In Cree, oftentimes it means they're missing something in their spirit or their family or their well-being. I sometimes think this concept of poverty can't even be translated properly, because our concept of being poor is to not have relations or not know where you belong. I'm wondering if you ever think about these words—to be poor and to work—and how those words don't necessarily translate properly in our ways of knowing?

**Derek:** I agree entirely with what you're saying because I've also conceptualized that way of thinking. And I think it's part of the indoctrination that we go through early in our learning. You know the idea that we're poor. My mom told me, when she was young, and even when I was young, we had no running water in our house, so we had an outhouse and we heated our home with a wood-burning stove in the wintertime. And that's how we lived. My mom told me that we didn't even know we were poor until somebody told us we were. It's all contextual when it comes to financial resources. You know, you don't need financial resources if you can go out on the land and grow a garden, which my grandmother did for us when we were kids. We always had fresh vegetables from the garden. We always had lots of bannock. We always had meat available, lots of dried meat and dried fish. We always had lots of food. I recall there was no real need for money, but I was a kid, of course, so I kind of idealize these things over time, but poverty is contextual. I think it's in language where we connect poverty with jobs.

**Crystal:** This is what we're trying to do as Indigenous peoples; bring this knowledge to this just transition framework, and our ways of knowing, being, and doing as Indigenous peoples. And language has everything to do with that. The literal translation of "kitimâkis," which Angele mentioned, is to be poor. But when you describe that translation, it has nothing to do with material wealth. When you say that to somebody, you're talking about what's missing in their life. Derek, you talked about being able to hunt and being able to grow your own food. When we talk about that word, that's the context that that word is used in. We actually don't have a word in the Cree language to talk about material wealth, because there's no such thing. And so I think that is an important conversation that's missing from this framework. And even when I participated in the writing of the Leap Manifesto, that was missing. And it was so hard to bring that conversation in.

I wonder, Derek, what you think? How do we bring these conversations forward? Because the just transition, unless we change the trajectory, as Indigenous peoples, this industry is going to be developed by non-Indigenous people yet again. And so, thinking about our ways of knowing, being, and doing, how can we bring that forward? How can we ensure that the just transition is inclusive, but also is grounded in biodiversity and healthy ecosystems like they say it's supposed to be?

**Derek:** Where it starts is in education. I think that for our young ones, like my kids, they have been coming with me to the Sundance ceremony ever since I started going back, a number of years now. It's a completely different type of education and learning environment when they're in Sundance, versus going to the classroom and learning through a provincial public education curriculum. For example, a lot of people think that Christmas is the time where you give things away. Canadian society is of the view that Christmas is when you give, but my kids know that Sundance in the summertime is when you give everything away. They've seen me give away everything of value, things that I'd received over the years as a Chief. Some of my most prized possessions that I actually had to cry to put them into my giveaway. So they know that these material things that we acquire are not permanent and that they're not a reflection of who we are. And sometimes the things we have are better in the hands of somebody else.

That's just one example; it starts with the young ones. It's going to take years for our young ones who are coming up observing these things, because I never observed it when I was a kid. We were still in the grips of our residential school aftermath in my family. I never saw the Sundance until I was a man, and now that these ceremonies are gaining strength, the message is getting out there more broadly. We're seeing, for example, education curricula that are now being built on land-based education. That, to me, is where a lot of this transition has to start, in informing the young people that there's other ways of living on the land, there's other ways of interacting with the land, that doesn't mean taking from it and these false notions of wealth or material wealth. We've got to figure out ways of eradicating our thinking of those things.

### Fighting for Universal Public Services

The conversation between Derek, Crystal, and Angele highlights Indigenous ways of knowing and understandings of economies of care. We need to rethink concepts of livelihood, working, wealth, and poverty, and we need to move towards caring, respectful economies that are not based on the accumulation of wealth, but on ensuring the necessities of life are provided for all, including through public services. Prominent sociologist Diane Elson highlights what a feminist economy of caring would be based on:

A caring and sustainable economy is based on mutual support and respect for rights. It is oriented to the broad and inclusive aim of improving our well-being in ways that reduce inequalities, not only today, but also for future generations. It prioritises care for people and for the planet.[26]

The vision Elson articulates is in line with Indigenous economies of sharing, caring, and reciprocity. Another way is possible, and it should be decolonial, feminist, and not capitalist.

Here we outline general programs and services worth fighting for, with the understanding that specifics will need to be negotiated in greater detail.

**Expanded healthcare:** Many people in so-called Canada are proud to have a universal healthcare system. And it is remarkable that a program that started in Saskatchewan in 1962, despite fierce opposition, is now Canada-wide. There is no charge for those covered by medicare to see a doctor and receive treatment. But there are, unfortunately, huge caveats to this, like if you need care for your teeth, eyes, or ears, your mental health, and much more. These other needs must also be covered by the public healthcare system, as they are in many other countries already. And healthcare must be available to all, regardless of immigration status and without discrimination based on race, weight, or other factors, as too often happens in practice. The experiences of healthcare workers as well as patients could also be much more enjoyable if we adequately fund the system, reversing decades of underfunding. In Saskatchewan, many people still remember having dental hygienists visit their public schools, decades ago, to do basic dental care. This was a relatively low-cost program that helped detect problems early. The program was eventually cancelled, and dentist associations tend to be opposed to public dental care, as they may make less money than in the current private system. Pharmaceutical companies are likewise opposed to publicly funded pharmacare. It is going to take strong popular movements to win these highly beneficial demands.

In March 2022, the federal Liberals and NDP joined a confidence-and-supply agreement, under which they would launch a new dental program for middle- and low-income families, to be fully implemented by 2025.[27] The agreement also includes provisions for a pharmacare system, starting with a Canada Pharmacare Act to be passed by the end of 2023. If implemented, these would be significant improvements to Canada's current public healthcare system; however, critics feel the implementation of the pharmacare program is stalling and moving too slowly with its open-ended deadlines, which are costing families billions in drug costs.[28]

**Universal education:** Imagine there were no longer financial barriers to you or anyone in your community accessing educational opportunities. This is already the case in a number of places in the world, including much of Europe, Cuba, and other places in Latin America. Fostering a care economy means fully supporting all of our educational pathways. And because we collectively have so much to learn to make a just transition happen, and to realign our economies to be in balance with nature and with each other, education is going to play a big role. We're all in this together, and our education system should reflect that. By drastically reducing tuition fees—or eliminating them altogether—and supporting people during their time as students, we can diminish student debt, a major barrier to education. This goes for colleges, universities, vocational skills training, and also continuing education for adults who are looking for new knowledge and skills that will be useful in a just transition. When education and training are more accessible, people will have access to decent work and incomes, which will allow them to live a good life (miyo-pimatisiwin).

**Childcare and eldercare:** Caring for children (miyo-ohpikinâwasowin) and caring for elders are very important parts of Indigenous economies. We can make this principle part of the just transition by making childcare and eldercare public and universally accessible. Various studies have shown countries that support integrated universal childcare policies have the potential for more economic prosperity and better work-life balance for parents, and that publicly funded universal childcare is of better quality than for-profit service.[29] Improving the quality, accessibility, and affordability of Canada's childcare system is important for supporting families, increasing women's participation in the workforce, and reducing poverty. We likewise need to prioritize good, public care for our elders. A green economy shouldn't just be about building more things, with construction jobs mainly held by men, and making profits. Creating a care-led recovery after the COVID-19 pandemic will create more jobs—including for women, who often do care work—and result in greater equity.[30]

**Publicly funded long-term care:** The pandemic made it abundantly clear that the for-profit long-term care system in Canada is completely inadequate, if not deadly. Residents are at risk, as are workers, who are often racialized women working at understaffed facilities in precarious contracts. Public and properly funded facilities can provide much better care and work conditions. Instead of being a way for corporate investors to pile on cash, long-term care can centre the needs of seniors. The Alberta Union of Provincial Employees, which represents 58,000 healthcare workers, is advocating for an increase in hours of care per resident to an adequate level, rather than below the bare minimum.[31] The union has also advocated for an increase in wages and benefits of all continuing care workers, to ensure paid sick leave so sick workers don't come to work, and proper personal protective equipment.[32]

**Public luxuries—libraries, community centres, and other public spaces:** There are many other programs and amenities that make our communities enjoyable and interesting places, and generate few emissions. As George Monbiot, long-time environmentalist and columnist for the *Guardian*, puts it, "public luxury available to all, or private luxury available to some: this is the choice we face."[33] Choosing public luxuries means we can have more and better-resourced libraries, community centres, parks, urban gardens, sports fields, theatres, and museums. In rural and reserve Indigenous communities, it could mean proper funding for language revitalization programs and land-based teachings, which often are done by volunteer elders or through underfunded schools. In urban Indigenous communities, it could mean ongoing permanent funding for Native Friendship Centres that run cultural and recreational programs like drumming, beading, and sports. Many jobs are involved in running these programs, and a strong commitment to this programming would create many permanent, local jobs. The key is to make sure the public luxuries are good services with decent, well-paying jobs, and that staff get the resources they need to run programs effectively.

### Ensuring Decent, Low-Carbon Work
Everyone here in so-called Canada and around the world should be able

to find decent work. But we know many people cannot, and that needs to change. The labour movement, including unions, has done a lot to improve working conditions, but employers and investors in a capitalist system have a strong incentive to push back those gains and exploit workers as much as possible. We need to work together to secure better labour regulations and conditions, and good public sector jobs and services.

There are many work conditions we can strive for. One is a shorter full-time workweek, like four days a week, which brings better quality of life, less need for frantic consumerism (as it frees up time to make things at home), as well as less emissions from commuting. Iceland is doing this, as is Between the Lines, the publisher of this book. The minimum wage, which in many places is a poverty wage, desperately needs to be raised, and everyone should be supported in old age, including through good pensions and good old age security. Temporary foreign workers and people with other forms of precarious immigration status need robust workers' rights and residency status in Canada, to dismantle the system of horrific discrimination, exploitation, and abandonment many face. And we should carefully consider how to provide robust income supports so no one lives in poverty, whether that is through increasing social assistance, disability support, ensuring full employment, or implementing some form of universal basic income. The key will be to do this in a way that truly benefits people—not just business interests—and strengthens public services at the same time.

### Clawing Back the Death Economy

One way to make space for Indigenous care economies to flourish is to dissolve the systems that have been preventing them from being enacted more widely. That the fossil fuel sector must be phased out is clear—but policing, prisons, the military, border security, and child welfare systems in Canada are even more directly extractive towards our communities and must also be abolished.[34] These "services" are not only expensive, but they also serve to prop up a warped and structurally racist interpretation of safety and security. They actively block work towards a just transition and care economy, as with the criminalization of land defenders, the theft of children from families, the violent protection of Canadian mining operations abroad, and the deportation, detention, and double standards at our borders. These "services" affect Indigenous, Black, impoverished, and racialized communities the most, but benefit only the wealthiest corporations and families. Shifting this funding and winding them down would go hand in hand with building new systems of safety and security that truly serve the public—including many of the universal public services we looked at above. It will also mean working with other nations towards equitable and peaceful international relations over time.

As the Building the World We Want collective puts it, defund the police organizing is

> part of a broader abolitionist vision for addressing the root causes of harm and violence in our societies instead of policing them. It is part of a movement to fund decent, long-term affordable housing, public transit, and community-led anti-violence projects rather than using mass public funds on forces whose primary role is to surveil, arrest, brutalize, incarcerate, and kill.[35]

We should work to phase out this harmful spending and retool the labour it employs and the resources it uses towards a globally just transition here, as well as to pay reparations for Canada's role in colonialism and climate change. This is not just a question of freeing up a wasteful use of resources, but of whether we are setting up a globally just transition that allows for self-determination of Indigenous Nations within Canada's borders as well as all nations and peoples outside of it.[36]

## WE TRULY ARE ALL RELATIONS
### Angele Alook

We began this chapter with Reyna, the live-in caregiver, evacuating during the 2016 Fort McMurray fire that devastated the oil and gas city. Prior to that fire was the big 2011 Slave Lake wildfire, during which seven thousand residents were evacuated to nearby Athabasca and Westlock. And in 2019, wildfires broke out near Slave Lake again, this time forcing the evacuation of the community of Wabasca and Bigstone Cree Nation for fifteen days. That whole summer in 2019, many First Nations, Metis Settlements, and rural communities were on alert. The interior of British Columbia experienced the same for much of the summer of 2021. What people from those communities remember is pulling together, Indigenous and settlers, rural and city folk; even the province and federal government seemed to pull together when they needed to.

When the huge Fort Mac fire happened, people were first evacuated north to Indigenous communities, because the roads leading south were blocked by the wildfire, then evacuated south to the city of Edmonton. It seems in northern Alberta, we take turns taking care of each other during and after the fires. My family in Wabasca remembers the Slave Lake fires and people coming to Wabasca for help, and the people of Slave Lake and Edmonton were there for us when Wabasca was evacuated. I remember my neighbours in Edmonton volunteering and donating items to the Fort McMurray evacuees. I even had family in Calgary taking in Fort Mac people. When the 2019 fire happened, my family lived with me for fifteen days in Edmonton, and although our human relations got out safely thanks to the community pulling together, my family relied on the Second Chance Animal Rescue Society (SCARS) charity to rescue our four dogs and the team of horses we left behind. It was our settler relations from Wabasca that called us to give us updates on our animals, and they gave us a call when the fire got close so they went in to rescue our horses.

I have some criticisms of the federal government and province, because I recall our elders not being properly evacuated from one of the fires, and young relations being made vulnerable when they were forced to evacuate to the Edmonton Convention Centre. However, having lived through all of these wildfires, I see that communities pull together, and that neighbours actually care about one another. What if this was how we approached public services? What if this was how we approached the climate crisis in general? Because when a fire or flood comes, it does not matter whether you are on reserve, in the city, or on a farm. A fire doesn't care if you are a human or a horse or a koala—remember the three billion animal relations harmed in the 2020 bush fires in Australia? At the moment the fire approaches, we realize how we truly are all relations.

There is a fire approaching, and we shouldn't be worried about the oil and gas infrastructure (Fort McMurray) or the impact on the forestry industry (Slave Lake) or any growth-economy industry; we need to build caring economies. We need good, decent jobs with livable wages, jobs in healthcare, childcare, education, eldercare, and

sustainable energy industries; we need affordable sustainable housing, we need food security. We need to centre our economies on
the lives of future generations, taking from the lessons of our elders.
We need to realize the earth is our mother as she gives us life. We
need to understand water is life and keep the oil in the ground so it
doesn't pollute the water. We need to show humility and respect for
all our human and non-human relations.

# 6.
# CHANGING THE
# POLITICAL WEATHER

A world where we uphold Indigenous sovereignty and build a just transition for all can feel far away. On a bad day it may seem impossible. But this future is well within reach. We can start by imagining what 2025 could look like if we start to pull together towards a decolonial just transition. We can imagine, for example, broad coalitions of settlers and newcomers standing in solidarity with Indigenous land defenders—to stop consent-violating projects like Coastal GasLink for good. Beaver Lake Cree Nation wins their constitutional challenge to show that the cumulative impacts of oil sands and other industrial development on their land are a Treaty right infringement, unlocking a legal precedent for other Indigenous peoples to use to protect, reclaim, and remediate their lands. Community alliances led by Black and Indigenous organizers win cuts to municipal police programs, like the removal of school police officers and an end to purchases of military-grade equipment. Workers join committees or form independent caucuses in their unions to build support for sector-specific transition plans—like auto workers in Oshawa advocating the General Motors plant be retooled as a publicly owned facility for building electric buses and delivery vans.

Youth climate strikers, union members, and grassroots Indigenous organizers in areas dependent on fossil fuel production hold town halls together to develop local just transition proposals. First Nations fighting new LNG export terminals on their land work with coastal communities in Indonesia and Philippines where this gas will be imported to stop these deadly plans on both sides of the Pacific Ocean. Local election candidates with strong ties to their community use people-powered outreach and bold platforms to win office. Metis Settlements in Alberta push the provincial government to uphold its funding commitments and build food gardens and solar projects. New organizations form to support union drives by migrant workers.

More academics, NGOs, and think tanks align themselves with mass movements by writing detailed policy proposals and helping provide popular education. Journalists and media makers expose the dangers of the old economy and connect the dots to show how different just transition efforts are connected. Settlers with access to land or excess money redistribute it to support new movement schools, Indigenous culture camps, and Workers' Action Centres.

The twin task of developing a shared vision of the future—what we have focused on so far in this book—is figuring out a shared roadmap for how we can get there. That is what this chapter and the next will focus on. The advantage we have over the powerful people defending the current death economy that seems so entrenched is that *everyone else* stands to live a healthier, happier, and more vibrant life if we build a decolonial just transition. Building stronger connections between us will make it possible to work together towards this shared interest. In other words, if we build dramatically stronger, larger, and more interconnected social movements like the ones starting to take hold in the vision of 2025 we sketched out above, we can unseat the current death economy and build a world where land and life can flourish.

There is a role for absolutely everyone in this work—and there is no shortage of skills needed, ways to get involved, and places to do it. (We will return to how you might get started or deepen your involvement in this work in chapter seven.) We are conceiving of social movements broadly here, using a definition of social movements that encompasses any collective efforts aligned with or explicitly working to achieve the vision we've laid out so far. Tactics like strikes, blockades, canvassing, and marches may most often be thought of as social movement activities, and these tactics are needed. But we also include tactics that use direct involvement in government processes such as legal challenges, political advocacy, and electoral engagement, as well as work to build our own alternatives, including community-owned renewable energy projects, Indigenous resurgence and land reclamation efforts, and workers' cooperatives. The bottom line is that the work needed to build collective power is far-ranging, and everyone is needed.

The seeds of many of the 2025 scenarios we described above have already been planted, with many people already working together to make them happen. Today's efforts are building on each other and paving the way for many more people to get involved, for smarter strategies to be created, and for a shared vision for a future like the one we have laid out to be built. If none or only some of this vision comes true by 2025, it does not mean it is too late. Our intention in sketching what a near-term shift could look like is to show that new political possibilities can be opened up quickly and that change often happens in a non-linear way. There is not a strict deadline after which hope is lost.

Today's movements are not yet strong or interconnected enough to force meaningful concessions from Canada's ruling colonial and capitalist political alignment, let alone replace it with an alternative one. But this can change. We can win the future we want by investing seriously in movement infrastructure that makes it easier for people to get involved and stay involved, rebuilding our skills for collective care and governance, and winning strategic fights that help us create a more favourable terrain for struggle. The question at the core of this chapter and the next might be best put by Paulo Freire, who once asked, "What can we do today, so that tomorrow we can do what we are unable to do today?" Freire was an educator and thinker whose work has helped shape landmark struggles across the globe—from anti-apartheid action in South Africa, the Landless Workers Movement (MST) in Brazil, land rights struggles in Gambia, and housing justice work in Scotland—and so he is a fitting mentor for this last section. We try to answer Freire's question in two parts. First, in this chapter, we look at what has been needed for recent collective action to mount a real challenge to power and identify some of the transformational wins we can organize for now that will unlock further possibilities. Then, in the final chapter, we ask how we can build coordinated collective power that can be sustained for the long haul and grown to the scale needed to win a decolonial just transition.

## What's behind Social Movements That "Win"?
Many of the changes we have proposed in the book so far are popular. Opinion polls are far from perfect barometers, but it is telling that 90 percent of people in Canada support a wealth tax and closing loopholes that allow tax havens, 79 percent would be more supportive of a transition away from fossil fuels during which financial support is provided to low-

and modest- income households, 72 percent support accelerated action to implement the calls of the Truth and Reconciliation Commission, and 51 percent support reducing the police budget in their municipality and investing the money in other city services. Even bold, direct tactics in support of Indigenous sovereignty and climate action are more popular than one might expect—for example, the cross-country Wet'suwet'en blockades in February 2020 had 40 percent of the public's support (even when they were polled with a leading question that didn't accurately represent Wet'suwet'en opposition to Coastal GasLink pipeline).[1]

But if these steps towards just transition are popular, why haven't they been taken? The shortest—and best—explanation is that it's because they are not in the interest of those with the most power in Canada (as we saw in chapter two). The most powerful actors in Canadian business, along with their backers in political parties and the media, work very hard to quell efforts for real action towards a just transition. They do so by constraining the public's imagination over what is possible, and by promising nice-sounding reforms that are only delivered in the shallowest ways. These tactics intensified under Canada's turn to neoliberalism in the 1980s and 1990s, and have continued to intensify in the years since. Governments and corporations have also stifled efforts from Indigenous Nations, unions, grassroots movements, or other groups to fight for a different world—both by applying laws that limit these groups' political activities and budgets, and through direct repression, collaborating with police forces as they need.

In a lot of popular culture and media, the work of social change is portrayed rather simplistically (if at all!)—if people get mad enough about a situation, they go out onto the streets and participate in a rally or march. But behind any visible wave of unrest, there is deep organizational, legal, and social infrastructure that helps social movements counter efforts to repress them, often laid years or even decades in advance.

In winter 2012, for example, a quarter of a million students in Quebec went on strike, holding nightly demonstrations, blockades, and massive marches that captured the public imagination, preventing a proposed near doubling of tuition fees. In the decade since, tuition increases have remained an untouchable issue for three successive provincial governments.[2] The Quebec student movement had had a ton of practice, with unlimited general strikes like the one in 2012 happening nine times since 1968, and more localized and time-limited strikes even

more frequently. This allowed for a regular cycle of mass political education and for student leaders to pass lessons and training on to new generations. Exercising this strike muscle so often helped defend a highly
organized and well-resourced student union network with mandatory
membership, direct democracy, and a legally protected dues-paying
structure.[3] Finally, because students vote on strikes and marches, and
the police crackdowns that usually follow, visibly disrupt daily life, the
student body and general public are forced to pick a side, reshaping the
political consciousness of many observers. Co-author Bronwen Tucker
was one of many students who started off 2012 on the sidelines, casu-
ally debating the merits of the strikes with her friends, before becoming
involved in the strikes as a participant and later an organizer. In many
cases, including Bronwen's, this led to long-term political involvement
in other struggles.

The February 2020 rail and highway blockades in solidarity with
Wet'suwet'en land defenders blocking the Coastal GasLink pipeline,
using the slogan #ShutDownCanada, were just one outcome of another
effective long-term movement building effort. Generations of land
defenders built the social, legal, and organizational foundations that
made it possible to freeze much of the transportation infrastructure
the Canadian economy relies on, bringing the reality of Indigenous
land rights to the public and delaying Coastal GasLink's construction.
A legal challenge in 1984—coming out of logging blockades on the
territory—led to the 1997 Delgamuukw Supreme Court ruling, which
recognized that Indigenous title is not extinguished in the areas claimed
by Wet'suwet'en and Gitxsan. Members of Unist'ot'en, a clan within the
Wet'suwet'en Nation, established a checkpoint and camp on a key area of
the territory to oppose Enbridge's Northern Gateway pipeline in 2009,
and regular action camps and a permanent Healing Centre built since
helped train and provide land-based healing to thousands of people
across Turtle Island, building a strong network. Before the fight to stop
Coastal GasLink began, this camp contributed to stopping Northern
Gateway, and helped spark and provide a model for other pipeline fights
like the Dakota Access Pipeline (Standing Rock), Energy East, Line 3,
Keystone XL, and Trans Mountain Expansion.

That collective action is ultimately behind most of the public goods
and systems of care that have existed throughout history is perhaps the
secret that capitalism has worked the hardest to hide. How we learn

about social movements is often warped because of this omission. We hear about the biggest moments and wins and losses, like when a pipeline is cancelled or built—but the impacts of social movements are often non-linear, like when a struggle in one place lays the groundwork for a future struggle elsewhere. Social movement history is often revised by those in power to serve their own ends. For example, legal and policy wins that were made possible by social movement organizing are frequently attributed solely to politicians or legal cases at the moment a paper is signed or a vote is passed. The media has largely portrayed the Occupy movement in the early 2010s as a failure because it is hard to trace direct results. But to take Occupy in New York as an example, huge swaths of the people who were first politicized there have gone on to be instrumental in many vital campaigns, from rent strikes to public healthcare advocacy to forming workers' co-ops.[4]

### Picking Winnable and Transformative Fights

As a framework to help think through where to focus our time and energy in order to open up a clearer pathway to a decolonial just transition, we look to André Gorz's concept of "non-reformist reforms." Gorz developed his idea in the 1960s in France while frustrated with the approaches of some unions and social democrats who believed capitalism could be "fixed" through the existing parliament and electoral system alone, and some Marxist groups who believed a revolution could be easily triggered in the existing political conditions.[5] His ideas have since been developed and refined by a huge range of movement actors, from Vancouver-based migrant justice organizer and writer Harsha Walia, to abolitionist organization Critical Resistance and the many other groups behind efforts to defund the police. Here we use adapted criteria from queer and trans liberation organizer Dean Spade and workshop curricula from PowerShift—a series of youth climate justice convergences held in so-called Canada from 2009 to 2019—to help identify transformative targets in the context of just transition and decolonization.[6]

When trying to evaluate a potential strategy for social change (big or small), you can ask a few key questions:

- Does it materially increase the strength or size of Indigenous jurisdiction, reduce emissions, and/or improve conditions for working-class people? (Ideal targets will do more than one of these at once!)

- Does it leave out the groups most impacted by an issue or further encroach on anyone's ability to meet their basic needs? (It shouldn't!)
- Does it legitimize or expand the power of the fossil fuel industry, corporations, or colonial government systems we are trying to dismantle? (It shouldn't!)
- Does it mobilize the most affected for an ongoing struggle by growing collective power, freeing up resources for movements, and/or making the vision of a just transition more tangible? (It definitely should!)

These questions can be summarized as: Is the strategy likely to improve material conditions, avoid harm, undermine corporate or colonial power, and provide advantages for future struggles? In other words, *Is it transformative?*

Alongside this test of whether a strategy could be transformative, you can apply a second test: Can you map out feasible pathways to securing your goal? In other words, *Is it winnable?*

Let's test some strategies using our framework. First, consider supporting a proposal to build a blue hydrogen facility where a First Nations band holds a majority ownership stake. Blue hydrogen, as we saw in chapter three, is a new favourite "climate delayism" tool of the fossil fuel industry. It involves using fossil gas to make hydrogen fuel, and expecting to bury the emissions created by this process using carbon capture and storage. Based on current Canadian government positions, we can assume this proposal is probably winnable. But as may be easy to predict, this is not a non-reformist reform because, even though it could potentially redistribute some wealth to an Indigenous Nation in the short term, it lives fully within existing corporate and colonial governance models without challenging them. Such a project would legitimize corporate and colonial governance models, and help prolong the ongoing operation of the fossil fuel industry by allowing it to appear climate and environmentally friendly.

A more complicated case is a fossil fuel divestment campaign aimed at a major Canadian bank. This strategy has the potential to pass our framework test, depending on how it is pursued. We can see first that a bank ending its fossil fuels investments is likely to materially reduce emissions by raising the cost of capital for fossil fuels projects, meaning fewer will be built. Second, if ending investments that violate Indigenous

rights is a core demand and is done in partnership with Indigenous organizers, it includes those most marginalized by the bank's fossil fuel activities. The third question is where things become a bit more complicated.
Bank divestment is already working to erode the fossil fuel industry's
power, but we would argue that its potential to erode corporate and
colonial power is much higher if it is tied to a longer-term strategy for
financial institution reform.[7] Investments by banks, along with other
large institutional investors, play a significant role in planning our economy. This function needs to be brought to far more democratic ends.
Otherwise, banks will be able to continue with current levels of profit-
seeking, speculation, and resource consumption that are incompatible
with ecological limits or providing basic needs for all in the long run.
For example, they might be incentivized to move to harmful investments
that are simply more favoured by the green capitalism of the moment,
like lithium mining for electric vehicles or financializing housing to make
it even less affordable. And more directly, private hedge funds or other
financiers are likely to pick up some of the bank's discarded fossil fuel
projects.[8]

If a bank divestment campaign is used to get new, previously unen-
gaged people involved—like the youth-led group Banking on a Better
Future is doing—and the work is used to make a vision of public banking
outside of capitalism more tangible to both organizers and the public,
then it can pass the last test too. So, with some care a private bank divestment campaign can be transformative. We can say it is also likely winnable, because Canadian banks are lagging behind even their international
peers on limiting these investments, providing powerful leverage.[9]

What follows here are four transformative and winnable strategies
we have identified as particularly fruitful levers for putting bigger wins
towards a decolonial just transition within reach. We offer this framework in the hope that it can be generative in the face of the sometimes
overwhelming scale of the challenges we face. But we do not mean to
suggest that "perfect" strategies exist. Political struggle is always going
to be messy and imperfect; the best we can do is to be anchored in the
future we want to win.

### All in for Land Back

Let's start with one group of strategies that are very much underway.
Efforts to assert Indigenous sovereignty and jurisdiction are not at all

new; they have existed since colonizers first came here. These efforts have resurged since the fight to stop the Northern Gateway pipeline; Idle No More and Standing Rock are two prominent manifestations. In recent years, these efforts have been popularized by Indigenous meme-makers and land defenders under the banner of "Land Back." And as we've asserted throughout this book, we can't get to a just climate future without land rights wins.

The Yellowhead Institute's *Land Back* report names three main strategies for reclaiming land: (1) environmental assessment and monitoring that involve affected community members and can delay or block resource projects like pipelines that threaten land and life; (2) establishing consent protocols and permitting developed and enforced by Indigenous Nations; and (3) physically reoccupying the land to disrupt Canadian jurisdiction and provide services to the community.[10] So, Land Back initiatives go beyond efforts like the Wet'suwet'en Nation's work to stop Coastal GasLink pipeline or the Tiny House Warriors in Secwepemc territory blocking the Trans Mountain Expansion pipeline. They include community-based environmental monitoring programs like the one that studies fish health in Beaver Lake Cree Nation. Culture and language camps, like the Nimkii Aazhibikong camp that is reoccupying Anishinaabe land on the north shore of Lake Huron, are another Land Back effort. Nimkii Aazhibikong is "Anishinaabe people, doing Anishinaabe things, on Anishinaabe lands."[11] This work helps connect young people with elders for cultural and language teachings and facilitates cultural resurgence of traditional Indigenous land and resource protection and management.

The *Land Back* report also names legal challenges as a potential strategy—though with caution, as this strategy comes with the risk of damaging legal precedents that can then be applied in other struggles. For every landmark win like the Delgamuukw Supreme Court case, there's a risk of industry and government wins that undermine the Canadian case law in support of greater Indigenous land rights recognition. This was the case with the Chippewa of the Thames First Nation losing their case in July 2017 objecting to the Enbridge Line 9 pipeline from Sarnia to Montreal, and the February 2020 Federal Court of Appeal ruling in favour of the Trans Mountain Expansion pipeline.

These various strategies are most likely to win if they're used in combination, providing both the political support and the formal legal power to make real gains. They're also far more likely to win if non-Indigenous communities get involved and forge meaningful solidarity. This solidarity can take many forms. One of the simplest ways to help is to learn what efforts to uphold Indigenous sovereignty are happening closest to you and find out or ask directly what is needed. In cities, Indigenous-led mutual aid efforts are occupying culturally significant sites in order to also provide housing and services for houseless community members. This was the case with Camp Pekiwewin in Edmonton (amiskwaci-wâskahikan), which lasted 105 days in the summer of 2020, providing harm reduction to prevent overdoses, hot meals, and pharmacy services, and won some concessions including more public shelter spaces.* This effort was able to endure because of the support of previously uninvolved community members, such as delegations from faith groups delivering scheduled batches of hot food and supplies, tradespeople building temporary shelters, housing outreach workers matching people with housing, and many others donating supplies as needed.

Those with any access to wealthier settler networks should work to redistribute land and resources (we'll talk more about this kind of redistribution in chapter seven). The RAVEN Legal Trust is doing this kind of work, raising legal defence funds for Indigenous Nations to assert their rights and title to protect their traditional territories.[12] RAVEN has backed many landmark cases, including the Beaver Lake Cree case in Crystal Lameman's Nation and the 2018 *Tsleil-Waututh v. Canada* case that "quashed" the Trans Mountain Expansion pipeline for a few years. Another form of leverage you might hold is in your workplace. For example, in February 2021, "a sizeable minority" of Baffinland mine workers stranded because of a Nuluujaat Land Guardians blockade of the mine site released a statement in support of the blockade:

> We recognize the Inuit as the rightful custodians of this land, and as the people who should make the decisions about how it is used. . . . This country has seen the consequences of entitlement and greed that have led to the destruction of the land for profit, and we are glad you are fighting for autonomy over your land. You've said that it is not the workers you are upset with, but the Baffinland executives, and we

---

* As with many collective struggles, these victories were partial. The use of the land in question and many of the demands of the camp remain unsettled at the time of writing.

would like to say that our support is also not with our superiors in the company, but with you.[13]

If these Land Back efforts to recognize Indigenous sovereignty continue to grow and gain cross-movement support, what could it mean a few years from now? Most directly, industrial projects that Indigenous peoples are currently fighting, like proposed LNG terminals on the west coast, will likely be cancelled for good. We could see precedent-setting court wins and government concessions to recognize full jurisdiction of important parcels of land, especially public lands currently claimed by the Crown. And as safer and more accessible gathering places to share ancestral cultural practices are opened and defended, these efforts could create a sea change in language revitalization, rebuilding food sovereignty, and land-based healing. These efforts would also build climate resilience by allowing Indigenous stewardship to recommence, more resources for Nations to own and operate housing and energy projects, further legal protections to stop future unwanted developments, and the renewal of Indigenous legal and governance systems independent from the Canadian state.[14] These are all critical building blocks for the decolonial just transition we need to see.

### Defund the Bad, Refund Land and Life

Our description in chapter two of the neoliberal austerity that has dominated Canada's politics since the 1990s glossed over a key point. Canadian neoliberalism *has* ushered in an era of austerity, but it has only resulted in budget cuts for working people and the environment—borne most disproportionately by Black, Indigenous, and other racialized people and communities that are made vulnerable in our society. Fossil fuel companies, the military, police, large corporations, and the wealthiest families have all actually received more support from the government. We have starved public goods, land, and life in order to feed Big Oil, corporate profits, and the security that capitalist growth requires.

Questions of "But how will we pay for it?" are easiest to combat when you can point to tangible (and truly giant) expenditures that actively harm our communities. The answer is that there are plenty of options: taxes on high earners and polluting firms, cutting military expenditure, long-term investment in green infrastructure, to name a few. The real issue is political will and political power. Just think about the impressive government policies put in place in the span of weeks when the COVID-19 pandemic first hit. This crisis has shown us that, when there is political will, the money and policy tools are there. And it is worth pointing out to anyone who asks this question that what we do not spend on climate action, adaptation, and upholding Indigenous sovereignty today will make this work much more expensive later on!

Including the largely invisible flows of public money as part of what's up for public debate helps to open up an accessible and potentially transformative conversation about what we could build instead. By asking tangibly what it would look like for the police to have less power over our communities—and particularly Black and Indigenous communities—we can start a public conversation about imagining and building a truly safe world.

The "refund" part of this strategy would include supporting many of the solutions we covered earlier, from universal public transit, to direct Treaty-based funding for Indigenous Nations, to affordable energy-efficient public housing, to community-owned renewable energy, to Canada forgiving illegitimate debts and paying reparations abroad to make space for a globally just transition. The exact demands can and should be made more specific to communities as they organize. In most of these cases, as we phase out funding for programs that are *not* serving communities, there are also other programs that will need to be built up at the same time. For example, we need mental health support and public housing alongside the defunding of police and prisons (as many abolitionist thinkers have sketched in more detail than we can provide here).[15]

Some "Defund—Refund" wins have already come as a result of the historic widespread support of the Black Lives Matter movement in 2020. Twenty US cities saw successful organizing in 2020 to shrink police budgets by a combined $840 million a year, with at least $160 million shifted to community supports like mobile mental health teams, affordable housing, eviction protection, youth jobs programs, and increasing the minimum wage for precarious city workers.[16] These small shifts are previews of what stronger social movements could win. In Philippines, Indonesia, Ghana, and Morocco, governments have phased out some fossil fuel subsidies and used the available funds for cash transfers and social supports like education and health insurance for low-income households.[17]

The following is a non-exhaustive list of $180 billion a year in public money in so-called Canada that could be cut, shifted, and phased out to lessen harm and free up both money and the public imagination towards a decolonial and just transition. Winning even one-quarter of this amount in the next few years would free up more than five times what the federal government was planning to spend each year on climate-related infrastructure and programs as of 2021.[18] These figures are taken from a 2017 to 2019 average where possible to avoid potential anomalies in government spending during the beginning of the COVID-19 pandemic; exceptions noted. For context, in 2019 the federal, provincial, and municipal governments together spent a total of $750 billion a year.

## But how will we pay for it?

Five harmful areas of government spending we can defund to free up $180 billion a year to refund land and life

    A.  Police, prisons, and border security: $23 billion
    B.  Military: $22 billion
    C.  Fossil fuel subsidies: $24 billion
    D.  Taxing the rich and corporations: $98 billion
    E.  Highway and aviation expansion: $13 billion

**A. Police, prisons, and border security: $23 billion a year.** Canada spends $15.7 billion annually on municipal, provincial, and federal police services;[19] in addition, $2.2 billion a year on the Canadian Border Services Agency, $5 billion on prisons, and $500 million on the Canadian Security Intelligence Service.[20] Adding it all up, about $23 billion a year. Policing, prisons, and border security prop up a warped and structurally racist interpretation of safety and security.[21] These institutions are actively blocking a just transition, as with the criminalization of land defenders. Shifting this funding and winding down policing and prisons would go hand in hand with building new systems of safety and security that truly serve the public.[22]

**B. Military: $22 billion a year.** It costs $22 billion a year to operate Canada's military.[23] There are also large capital purchases like the estimated $19 billion for fighter jets Trudeau has been delaying since he entered office.[24] Clawing back this spending will free up a wasteful use of resources and allow for self-determination of not just Indigenous Nations within Canada's borders but also all nations and peoples outside of it.[25] As with defunding the police, phasing out military spending means re-envisioning safety and security and building up new public programs that truly deliver it. It also means working with other nations towards equitable, peaceful international relations.

**C. Fossil fuel subsidies: $10 billion a year in direct subsidies, $14 billion a year in government-backed public finance.** As we saw in chapter three, the government's financial support for the fossil fuel sector in Canada is huge and is prolonging the fossil fuel era— allowing new oil and gas projects to be built that otherwise would not see the light of day. The "direct subsidies" here include the average of known tax breaks and direct transfers for specific programs and projects for years 2017–2019. The "public finance" bucket is almost all preferential loans or guarantees through Export Development Canada, which also means liability for any projects that become stranded assets will be held by the public. Subsidy totals fluctuate greatly year to year due to commodity prices and one-off programs, and it is important to note there are also large portions of subsidies, including tariff exemptions and clean-up costs, that are easily and frequently avoided by companies, where there is not enough public information to calculate them. These figures do not capture some large new subsidies since 2020 like Alberta's finance for Keystone XL or the new CCUS tax credit.[26]

D. **Taxing the rich and corporations: $98 billion a year.** As we identified in chapter four, the top 1 percent of people hold almost 26 percent of wealth in Canada, and this imbalance got worse during the pandemic. The $98 billion here comes from a tax on extreme wealth starting at 1 percent on net worth above $10 million (Alex Hemingway at the CCPA estimates this would raise $36 billion a year) as well as inheritance taxes, increased corporate taxes, ending tax breaks on investment income, and measures to make it harder to use tax havens (estimated at $62 billion a year by Canadians for Tax Fairness). Corporations and the rich have awful amounts of resources to throw at avoiding their money being redistributed, paying accountants and lawyers to help them avoid and evade taxes. So, it will take time and focused effort to build the political and regulatory power to adequately redistribute resources from corporations and the rich. We use moderate estimates here to estimate what might be possible in the near term. Others have rightly proposed higher rates of taxation as well as ceilings on maximum wealth, after which everything is expropriated.[27]

E. **Highway and aviation expansion: at least $13 billion a year.** Building new highways increases our dependency on individual passenger vehicles and is not even effective at decreasing congestion. It is also a huge public health hazard for air pollution and traffic accidents. Similarly, government handouts and preferential loans to airports, airline companies, and aircraft manufacturers are locking us into this fossil-intensive sector instead of building long-distance public transit and other options to reduce the need for flights. The public spending is often project-based and not well documented as a yearly average, so this is a likely underestimate based on 2020 and 2021 data in the Energy Policy Tracker, a coalition research project our co-author Bronwen Tucker worked on that is led by the International Institute for Sustainable Development.[28]

### Building Unions for the Decolonial Just Transition

Unions will be key to shifting the political terrain and winning a decolonial just transition in so-called Canada. Outside of the inherent land rights that Indigenous Nations hold and assert, unions are perhaps the most powerful tool we have to directly challenge capital and colonial powers. Unions are designed to facilitate workers acting collectively and democratically to achieve shared interests. They can pool resources towards their aims, run campaigns, bargain for demands, and exact costs from those in power by a coordinated withholding of labour through strikes, work stoppages, and other tactics. Even after the past few decades of neoliberalism, unions are among the strongest collective organizations available to exercise power—both for more fair workplaces and for the kinds of social policies and supports that are needed more broadly in a just transition.

It is because of the labour movement that many workers today enjoy basic rights like the forty-hour workweek and the eight-hour workday, parental leave, employment insurance, workers' compensation, sick leave, vacation time, compassionate care leave, domestic violence leave, rights against physical harm, and most recently rights against psychological harm in the workplace. This work is, of course, incomplete. Our vision for a just transition includes ensuring that fair wages, benefits, and healthy, safe working conditions that support a dignified life are available to *all* workers, including migrant workers regardless of immigration status, workers with Indian status, and part-time and gig workers.

Even more foundationally, unions uphold the right to freedom of association, protecting the related rights of collective action and collective bargaining. Unions have also been crucial to winning broader social goods, from employment insurance to public healthcare. Now unions can mobilize to achieve an energy transition aligned with 1.5°C and especially one that is planned and just. In other words, unions can provide both the political power to win the needed pace and scale of emissions reductions, as well as help ensure the reductions happen in a way that creates good, green jobs and other public goods. Without union involvement, we are much more likely to land on the path of mainstream "net zero" plans that we laid out in chapter two, where climate delayism and market-based solutions simply amplify our existing inequalities.

If we can build a bolder, bigger labour movement, it will unlock many tools and win cross-cutting reforms that would give all movements working towards a just transition better odds. First, it would help workers better defend and expand their rights in line with the good, green jobs for all that a just transition demands. It would also make unions more likely and better resourced to act in material solidarity with Indigenous peoples asserting their sovereignty. For example, Indigenous and non-Indigenous workers and Indigenous Nations could unite in campaigns, as was done in the 1960s with deadly arsenic emissions from Giant Mine in Yellowknife,[29] or unions could refuse to work on projects that do not have full free, prior, and informed consent of Indigenous communities.[30]

A larger labour movement more oriented to broader community struggles would also counteract the gravitational pull political parties feel to stick to neoliberal, colonial, and capitalist frames and solutions. It would push all parties, but especially the NDP, to take positions and actions in line with a decolonial just transition. If successful at a large scale, it would open up the possibility for powerful tools and structural changes—like general strikes, sectoral bargaining, and a shift to a four-day workweek—that are currently out of reach. General strikes are when a substantial proportion of the labour force in a city (or larger jurisdiction) takes part in joint strike actions, shutting down workplaces, often with cross-cutting and large-scale demands like reversing cuts to social programs.[31] Sectoral bargaining is an alternative form of labour law—in place in much of Europe—that lets workers in an industry or region share a collective bargaining process and ensure that similar work receives similar pay, benefits, and protections.[32] Winning the right to use sectoral bargaining in Canada would rapidly grow the portion of workers belonging to a union, and would also be a powerful tool for negotiating sector-specific decarbonization policies that are just and uphold land rights.[33] Lastly, a four-day workweek, alongside higher wages, would lower emissions and free up workers' time to rest, to build community, to play, and to fight for further wins—as we covered in chapter five.

But currently, these changes are out of reach. And in the case of sectoral bargaining and four-day workweeks, they risk leaving many of the most precarious, low-wage workers behind if pursued by the labour movement as it exists today. Too few of those workers' workplaces are unionized. To win a just transition, precarious and low-wage workers cannot be under-represented in the labour movement.

One problem is that as union activities have faced greater pushback under neoliberalism in the last few decades, many unions have focused on a narrow conception of the interests of their existing members, prioritizing workplace-level bargaining, fundraising for the NDP, and legal processes led by experts behind closed doors over organizing new workplaces and mobilizing for larger structural changes at their workplaces and in the wider world. One symptom of this trend is the massive decline in strike actions taken in Canada in recent decades—falling from about one day of work missed to take part in collective action per worker each year at its peak in the mid-1970s to one day for every *ten* workers for the past two decades.[34]

Here are two strategies (already underway in some places!) to help build a larger labour movement that is galvanized to fight for a just transition:

**Union drives led by precarious, low-wage workers:** Any new union drives will help build a bolder, bigger labour movement in Canada, but efforts are especially needed to unionize low-paid, precarious workplaces such as logistics, meatpacking, retail, hospitality, the gig economy, and custodial work. Workers in these fields are disproportionately Black, Indigenous, or racialized, women or non-gender conforming, and/or disabled; they should be at the forefront of the struggle for a just transition.

Even a few new bargaining units in these sectors would bring much-needed material benefits to precarious workers, and create momentum and legal precedents to make further wins possible. Union drives are also excellent crash courses in the tools needed for collective action for the long haul. If more precarious workers gain experience using unions as a tool, the labour movement at large can better be pushed to fight for them and their communities' needs beyond working conditions—for example, on environmental racism, migrant justice, and strong public services. There is already momentum on this front. Precarious young workers, racialized workers, and other equity-seeking groups are leading major changes in the labour movement.[35] Gig Workers United won the right in 2020 for workers employed through apps like Uber to unionize in Ontario, and they are building on this win to fight for full worker status.[36] In the face of the worsening working conditions and growing inequalities triggered by the pandemic, there are new organizing drives at Canadian locations of Amazon, Staples, Indigo and Chapters, Starbucks, long-term care homes, and more.[37]

We should be absolutely clear that these current union drives in precarious sectors led by Indigenous and racialized workers are part of a long legacy of BIPOC labour movement leadership across all sectors. Sikh sawmill workers and Indigenous shipyard workers in British Columbia were among the first unionized workers, and today many racialized workers are key organizers in public sector unions that represent care workers in hospitals and care homes (who often come as temporary workers from Philippines and Jamaica or new immigrants from Somalia). It has been mainly racialized women leading the Fight for 15 and Fairness movement (now called Justice for Workers!) to increase the minimum wage across the country. In Cold Lake, Indigenous women care workers led a forty-day protest in 2017 of a lockout in a Points West Living for-profit care home, fighting for better working conditions, increased staffing, and better care for patients. Also in large national unions, Indigenous and racialized workers have begun to organize and make their demands known to decolonize and confront the history of racism and exclusion in unions.

As we have seen over and over during the pandemic, many of the most precarious jobs are precisely those most needed for the economy to function day to day. There is an urgent need to ensure precarious workers receive the pay and protections they deserve. And further, these workplaces also have strategic importance for the labour movement as a whole to exercise its full potential towards a just transition.

**Bargaining for the Common Good:** Bargaining for the Common Good (BCG) is an organizing approach for bringing unions and community partners together to push for broader-than-typical bargaining demands in their shared interest. In an article about work refusals as a tactic employed in the pandemic, Nigel Barriffe, a local officer of the Elementary Teachers' Federation of Ontario and an organizer with the grassroots group Ontario Education Workers United, argues for the need for educators to engage with the community's struggles for justice outside of school: "We teach their children! Their children come to our classrooms! How could we not be making sure that we're staying with them when they say that they're [dealing with] a bad landlord?"[38] He explains, "In my local [union] . . . we're currently in a struggle for what I feel is the soul of our organization." Barriffe and his allies are fighting to orient towards BCG demands rather than staying in a "business unionism" model that just manages economic transactions and reduces friction between workers and bosses.

This kind of struggle for the soul of unions is happening across more than just one organization. The most prominent recent examples of BCG in action have been in US-based teachers' strikes that fought against budget cuts to schools beyond teachers' direct wages—from Oklahoma to Arizona to Illinois. The 2018 West Virginia teachers' strike was a particularly exciting example of BCG that connected working conditions, energy transition, and the care economy all at once—teachers fought and won the elimination of a tax break given to the coal and oil production in the state to help pay for an increase in wages and school spending.[39] But similar efforts exist here too: the Work in a Warming World (W3) research program at York University has compiled a database of existing clauses from Canadian collective agreements that aim to make their workplaces or communities more sustainable.[40]

One example of applying BCG approaches to the fight for a just transition specifically is the use of contract bargaining to push for retooling a struggling auto factory to build electric vehicles for Canada Post or public transit instead. Green Jobs Oshawa, a grassroots coalition of workers and community members, has been advocating since 2018 to make this change at the Oshawa Assembly facility. As we write this book, co-author Angele Alook is also part of the bargaining team for her university faculty association, where the team has been negotiating increasing equity provisions by increasing Indigenous and Black faculty hires, as well as attempting to write provisions to confront the climate change crisis into the collective agreement. In 2018 at the AUPE convention, Angele worked with the Environmental Committee as a staff adviser to include, for the first time, language in the union's constitution supporting the just transition of workers in Alberta. This was a big step for one of the largest public sector unions in the province. More and more public sector and private sector unions are writing language into their collective agreements and constitutions recognizing that our labour forces will need to transition away from fossil fuels.

There are many other exciting possible formations for BCG to be used for a just transition. On Land Back, we can imagine a public sector union, as part of their collective bargaining with government employers, calling for a tract of land that's been at the centre of wider community struggle to be put back under meaningful Indigenous control. Taking the mutual aid reoccupation of Camp Pekiwewin as an example, imagine if City of Edmonton workers were to bargain for this historically significant site to be returned to the governance of local Indigenous Nations, in solidarity with the camp organizers' public demands.

A BCG approach could also unlock just transition planning and support in fossil fuel–dependent communities. For example, bargaining demands could include government support programs to reach all affected community members rather than just those directly employed in the sector (who are on average more likely to be white male settlers and more highly compensated than workers in service or other "indirect" sectors oil and gas companies rely on to operate). As well, broader clean-up, health, and economic diversification issues that affect whole communities near fossil fuel facilities need to be redressed as part of a phase-out. It is nearly always Indigenous communities that have borne the greatest brunt of these impacts. If community groups, Indigenous Nations, and union locals in these regions start to form lasting relationships and develop joint priorities for a just transition to work on together, this kind of holistic planning will be much more winnable. Community-union-Nation visioning and collaboration can also go a long way to combat the rise of xenophobic, often white supremacist linked organizing to "defend" the fossil fuel industry that industry and right-wing parties have been carefully stoking in many fossil fuel producing regions.[41] The best way to deflate this rising far-right extremism that puts Indigenous, racialized, LGBTQ2S+, and many other communities in danger is organizing to show that a locally tailored just transition is more viable than continuing to rely on volatile fossil fuel economies.[42]

How you can support these strategies for a bolder, stronger labour movement depends somewhat on where you are situated in the economy. For non-unionized workers, look into unionizing your workplace by speaking to an organizer at a union active in your sector or reaching out to projects like Gig Workers United, the Canadian Freelance Union, chapters of the Industrial Workers of the World, or Unions Are Essential that can provide resources and advice.[43] For workers in existing unions, you can help scale up efforts towards these strategies by joining worker committees, forming an independent caucus within your union to push for political changes, organizing to win resolutions to guide union activities, and taking labour and mobilization courses through your union's education department.[44] If you are not working, not interested in union involvement at your place of work, or not able to join a union, you can still help by showing up to support other unions' efforts like information pickets or strikes.

### Winning Local, Public Goods

One more strategy that can be particularly transformative is organizing to bring utilities, housing, and other infrastructure or services into public ownership and democratic control—whether by Indigenous Nations, governments, cooperatives, or hybrids of these. This strategy can also include transforming existing public utilities to be more democratic and accountable to Indigenous Nations. Many parts of the vision we've set out for a just transition fit under this umbrella—from free and expansive public transit, to building Indigenous-owned renewable energy cooperatives, to winning public daycare and eldercare.

If we win public ownership of some services and infrastructure in the next few years, it will unlock resources and services to make working people's lives easier as well as emissions cuts that get us closer to a just transition. These local goods and services are often the most visible and direct influence of government and corporations on people's daily lives, so this is an excellent way to organize new groups and individuals into social movements. But these campaigns could be even more transformative if they begin a devolvement of power and governance away from for-profit entities to more democratic and locally accountable ones.

By removing the profit motive from the central goods and services that provide people's basic needs, we will be better able to ensure equitable emissions cuts are made on time and that people's needs are actually met. As you can recall from chapters three and four, we don't mean to say that centralized control or decision making is always misaligned with a just transition, but that overall, if Indigenous sovereignty is fully upheld, this will likely mean much more local and overlapping jurisdiction and governance. Local decision making will also help these goods and services better fit local needs and be accountable to the public. Lastly, if we are successful in building powerful campaigns and movements capable of winning local public goods that respect Indigenous sovereignty, we will have begun to practise the kind of good, local governance that should be a centrepiece of a decolonial and climate-safe world.

There are many examples of this strategy already in practice. The Vancouver Tenants Union is organizing for ten thousand units of afford-

able public housing to be built every year to fix houselessness and housing precarity. Tŝilhqot'in Nation fought to have their inherent rights to
self-governance affirmed by the Supreme Court in 2014, and now has
a permit system in place for all non-Tŝilhqot'in people wishing to harvest morel mushrooms on Tŝilhqot'in land that raises money for the
Nation's services. The Canadian Union for Postal Workers' Delivering
Community Power campaign aims to transform Canada's public postal
service network to be a tool for a just transition—including electrifying
their fleet, building electric vehicle charging stations at post offices, offering public banking and digital services, and providing elder check-ins.
SENTRO, a trade union alliance in the Philippines, has since the early
2000s been organizing workers and community members who had been
priced out of electricity access by the privatization of publicly owned
utility cooperatives. They have prevented further privatizations and created some pilot renewable energy projects themselves.[45] Walpole Island
First Nation and Sogorea Te' operate land trusts collecting "land taxes"
to finance the purchase of land title to put it back under Indigenous
control. Horizon Ottawa is working to keep respite centres that provide
essential services to houseless community members open and bring
privatized transit lines back into public ownership. There are campaigns
for free transit in Edmonton, Ottawa, and Toronto, among other cities,
linked and supported in part by Amalgamated Transit Union Canada
and the Keep Transit Moving coalition.

   It is incredibly difficult to create winning social movement strategies—by nature these struggles face long odds, because they mean going
up against institutions with consolidated power. We offer this framework
for identifying transformative and winnable strategies as a starting point,
not an end point, for developing a more detailed road map for overriding
this power imbalance once and for all for social change. If we win even
some of the examples set out here in the next few years, a decolonial just
transition could be within reach.

# 7.
# SIHTOSKÂTOWIN

## Pulling Together for a Just Transition

We will be a lot closer to upholding Indigenous sovereignty and building a just transition for all if we can achieve some "transformative and winnable" strategies, such as those we just covered, to help build bottom-up power and remove obstacles. But how can all of our collective efforts fit together to actually get there? To allow for a just and livable world to be born, we will need to build a new "political alignment" for a just transition that is strong enough to win real concessions from, and eventually unseat, the current political alignment that runs Canadian politics. In other words, we need to build a broad coalition of people with a shared vision of a just future and overlapping ideas on how to get there. We need this coalition to be strong enough that it becomes more costly for Canadian governments to continue on their destructive trajectory than to bend to our vision of a just transition.

In the longer run, we will also need to develop new ways of governing and sharing power in order to ensure decision makers truly represent our interests and needs. Drawing on and starting to practise living in the laws of miyo-ohpikinâwasowin (raising children in a good way), miyo-wîcihtowin (good relations and unity), sihtoskâtowin (pulling together for survival), and manâcihtâwin (respect and reciprocity with the land) will help us with each of these steps towards a decolonial just transition, as well as with living in the new world we build together for many generations to come.

This chapter lays out how we can build a new political alignment capable of enacting a just transition. One ingredient is winning mass support for this vision by using common-sense and compelling narratives that show the vast majority of people that our struggles are inter-connected and have common solutions. Second, we must be strategic about the ways in which elections and political parties can and can't help us. Lastly, we must deeply invest in "movement infrastructure" to make it possible for the huge number of people who stand to benefit from a decolonial just transition to work together to make it happen.

## A New Political Alignment for a Decolonial Just Transition

In the face of overlapping and intensifying crises like COVID-19 fallout, climate disasters, and the growing precarity of jobs and housing, it is easy to feel overwhelmed. But we can look to history for some good news: the breaks from the ordinary brought on by crises are historically when the largest structural changes in society have occurred. As powerful and embedded as our economy built on death seems, it is more fragile at this point than it has been in decades. Cracks are showing in the current dominant political alignment. Liberal and Conservative political parties have alternated in holding power and implementing neoliberal policies over the last few decades, supported by lobby groups like the Business Council of Canada, think tanks, and their allies in corporate media and in academia. But a growing number of people no longer see the major players as legitimate. Broken promises and austerity under the current neoliberal capitalist alignment are causing anger and desperation. Some of this dissatisfaction is being used to begin building our new political alignment. At the start of this decade, there has been more momentum in decolonial and liberatory social movements calling for a different way of doing things than most of us have seen in our lifetimes.

However, this polarization is also being harnessed by billionaires and others on the political right to further consolidate their grip on land and life. Growing numbers of working-class people, suffering the effects of job loss and the decline of social welfare, are being recruited into movements with white supremacist and nationalist features—often sup-ported directly or indirectly by right-wing political parties and corpora-tions. So, as intensifying climate change and income inequality destabi-lize more people's lives, there is both a new promise of a breakthrough towards a just transition and a real risk of an even more authoritarian capitalism taking hold. If we don't take advantage of present and future crises to build a new political alignment for a just transition, the far right may succeed in building a much more dangerous new alignment instead.

Which movements people support is, in part, an outcome of which stories they are told about what is happening in the world, and which ones they believe and see themselves as part of. A powerful story formula is what drives "populism." Put simply, populism casts a mass protagonist—the people—against a political antagonist—the elite. The problems and frustrations encountered by the people are explained as having their cause in the elite. In this story, the elite have betrayed the people. The solution proposed by the populist formula is the displacement or replacement of this elite by representatives of the people and their interests. We think a decolonial, left-wing populism could be a potentially powerful tool to build a political alignment that can win a just transition.

The right-wing populist view is that the elite have sold out everyday (white and male) working people and their values by accepting globalization and cosmopolitanism. These populists, thus, want to reinforce the power of white people in society (white supremacy), the primacy of private business (capitalism), and the dominance of their state over other states (imperialism). As Mi'kmaw legal scholar Pamela Palmater puts it, "Indigenous peoples have long had to resist the ever-present threat of racism, white supremacy, and anti-human-rights efforts—the issue we now face is that this racism has become very public or mainstream." This is in part due to the growth of right-wing populism. White supremacists have always been here, blaming the "other," but now their "hope is to cover their hatred under the guise of unity, nationalism, and public security, i.e. protection from the others who they claim pose these threats." Palmater describes how Indigenous peoples defending their traditional territories have increasingly become concerns of "public safety" in so-called Canada. "Indigenous peoples are experiencing a second wave of violent colonization by states and corporations, which are intent on developing Indigenous lands and resources."[1]

Right-wing populists appeal to racism, antisemitism, and conspiracy theories to explain their grievances about job loss, deindustrialization, and growing inequality. It is now common to see conservative politicians dog-whistle to these politics, riling their supporters up with racist and classist lies like the spectre of immigrants "taking Canadian jobs." The people mobilized by this messaging are not representative of the whole working class, although they aim to paint themselves this way. Looking at the supporters and participants of efforts like the anti-COVID-vaccine "Freedom Convoy" to Ottawa, Canadian Yellow Vests, United We Roll,

and the People's Party of Canada, they are disproportionately white and
feature owner-operators, landlords, and small business owners more
prominently than rank-and-file workers. Many of their leaders also have
been shown to have connections—whether formally or informally—to
well-resourced white supremacist groups.[2]

Left-wing populism, on the other hand, names a different elite. And it
has a strong potential to become the new, ambient "common sense" in our
society and mobilize the real working class towards a different political
alignment. In this story, the wealthy—a small group of capitalists, mainly
settler capitalists domestically and multinational investors abroad—are
understood to have steadily eroded the livelihoods of the people and
destroyed entire cultures. The solution involves displacing them with
more democratic systems and institutions and redistributing their wealth
to build a better, climate-safe world. The Occupy movement was a clear
expression of populism, casting the 99 percent of Main Street against the 1
percent of Wall Street. Jeremy Corbyn in the United Kingdom proclaimed
his Labour campaign "for the many, not the few"; Bernie Sanders's challenge to the office of the US presidency cast the "millionaires and billionaires" against the plight of working-class Americans.

In this sense, left populism can be used as a shortcut to left-wing
politics. The populist frame taps into an increasingly widespread discon-
tent and dissatisfaction by communicating problems in accessible and
clear ways that identify simple solutions. The left-wing version offers a
sensible explanation for why things are the way they are, and a tenable
solution for doing something about it. But we do have to avoid over-
simplifications and omissions. Several expressions of left populism have
paid insufficient attention to decolonization, painting the interests of the
masses as homogenous. It is possible to acknowledge different histories
and goals, while at the same time drawing parallels in order to present
a shared narrative of "us"—Indigenous peoples, workers, migrants and
displaced people, women, LGBTQ2S+, disabled, and young people—
and "them"—a small group of mostly settler capitalists. This is a powerful way to build a shared political alignment to counter the neoliberal
colonial status quo. And this initial populist frame is a starting point
from which broader networks of solidarity and deeper understandings
of racialized settler capitalism can form.

Populism is most often described as a tool to help capture or reform
governments, but in this case, the settler state has to be undone, not just

captured by a different "us." We think populism is a powerful tool that in the short run can help us do the former, but in the longer term can lead more people into deeper engagement with politics and struggle. In this way, left-wing populism can be a first step in building wide support for the kinds of localized, democratic, decolonized governance and ownership structures that we would ultimately want to establish through a just transition. If the solutions that are posed direct some people to get deeply involved in collective movements, a simple populist explanation can be an inroad to a more systemic societal transformation.

Right-wing populism and a decolonial left populism are not the only futures possible. The neoliberal capitalist order has shown an incredible capacity for withstanding and adapting to populist pressures. Political parties such as the Liberals in Canada and the Democrats in the United States have continued to shift their messaging to position themselves as the only capable managers of economic and social order. They roll out bold-sounding net zero plans that would actually lead to climate disaster and intense inequality, or make symbolic cabinet positions for racialized politicians rather than addressing systemic racism. They do so even as they oversee colonial violence and climate chaos. A new strain of neoliberalism may yet win out as the dominant political alignment—unless we build our own!

## What about Elections?

Rather than engage with the more complicated movement-building tactics outlined in chapter six, it's extremely tempting to say, "Let's just focus on electing the NDP (or Green Party)" or "Let's organize within the NDP (or Green Party) so that they will fight for a just transition." We certainly need all of the shortcuts we can get, but lessons from past efforts to push both of these parties towards a more "progressive" orientation show these are not promising strategies on their own. Instead, recent history in Canada and internationally shows that if any political party within neoliberal capitalism is going to be an effective vehicle to help win a decolonial just transition, they will need strong external social movements that can pressure them to adopt and follow through on helpful positions. This is especially the case when we consider that upholding Indigenous sovereignty means radically reforming the Canadian state as it exists today. We need to be clear-headed that it is fundamentally bottom-up pressure that is needed to secure this goal. We cannot rely on the government of Canada to keep a promise to undermine its own power.

However, this does not mean abandoning the electoral sphere entirely. We have a settler colonial government and a global capitalist economy, and too many decisions about distributing resources and social goods are held by national, provincial, and municipal governments to leave this arena of power to the forces in the centre and right of the political spectrum. Elections are currently how the majority of people in so-called Canada believe social change happens. They can and should be used as a moment of engagement to help bring new people into our movements, practise the skills of organizing, and develop political analysis.

The NDP was born as a merger between the labour movement and the socialist Co-operative Commonwealth Federation, at a time when political parties frequently housed and enabled many social movement functions. But with the turn to neoliberal capitalism in Canada, the NDP has become more vulnerable to pressure from corporations, media, and other political parties, and more concerned with public perception among the wealthier portions of the population that tend to be most engaged in electoral politics. It has largely abandoned the political education of its membership, the development of solutions to pressing societal problems, and advocacy for working-class concerns. As a result, when the NDP has held power provincially, it has not delivered bold or transformative changes.[3]

Another obstacle to the NDP being an effective vehicle for transformative change is the influence on the party of some of Canada's less progressive industrial unions. The labour movement in Canada is far from uniform. While parts of it have done much to advance a just transition as a policy framework and to stand in material solidarity with Indigenous Nations, other elements have blocked the NDP from taking such positions. At the 2021 federal NDP convention, many union delegates voted against and narrowly defeated a resolution amendment to oppose new fossil fuel projects and pursue a managed wind-down of production. Similarly, in the British Columbia NDP, pressure from some of the industrial unions has been a significant obstacle to the government ceasing to support old-growth logging and fracked gas expansion.

In the face of these shifts, many attempts have been made to push the NDP to be more democratic and to fight for the bolder positions its membership and the broader working class support (though until recently, few of those efforts have prioritized decolonization). In the late 1960s the Waffle, a group within the party, aimed to make the NDP over with anti-imperialist and socialist values. It saw some early enthusiasm and success, but was ultimately repressed by party leadership, due in large part to threats from a number of unions.[4] In 2001, an attempt to renew the party under the banner of the New Politics Initiative proposed dissolving the NDP and reconstituting it with Canada's progressive organizations and social movements as integral components. Most of this reformation effort failed, and the NDP instead centralized, professionalized, and modernized operations.[5] At the 2016 federal NDP convention, progressive activists pushed to have the party adopt the Leap Manifesto, which included among its positions fully upholding Indigenous rights and ending future fossil fuel infrastructure construction—earning itself immediate backlash from mainstream media and many within the party, particularly the Alberta NDP. The Leap Manifesto was ultimately adopted for discussion within individual electoral district associations only, allowing the central party to ignore it. Since 2017, activists with the Courage Coalition have been organizing at the party's conventions, pushing the NDP to adopt bolder policies. These initiatives are admirable and have likely helped prevent a deeper slide rightwards, but none have fundamentally transformed the NDP.

The NDP's modern market-driven messaging strategy requires that any news or communications generated by the party, its leaders, or its campaigns are targeted to particular segments of the population that could expand the party's base of support. This philosophy of message discipline leaves little interest in potentially negative or controversial news stories about internal debates on policy or strategy.[6] This effectively means, however, that many of the formalized places within the party where internal discord or debate could happen are now seen as risks to message discipline—and are therefore limited and much more controlled. Co-author Joël Laforest experienced many of these difficulties first-hand when campaigning for the Alberta NDP against Jason Kenney's United Conservative Party in 2019. Notley's term as premier of Alberta had seen the party become fixated on "winning" the public discourse about pipelines, leaving little space for more class-based politics to emerge—and often reinforcing conservative messaging about the importance and necessity of fossil fuels.

The bottom line is that transforming the NDP will require external pressure from social movements. Even if organizing to shift the NDP from the inside is successful, it will still face the full force of the settler colonial state and global capitalism. As labour studies professor David Camfield argues:

> A left-led NDP government would buckle under the pressure of its enemies—capitalists who'd refuse to invest, top state managers, the pressure of bond and currency markets—as SYRIZA did in Greece in 2015 *unless it was driven forward by a powerful social movement that it couldn't control.*[7]

Therefore, our first priority should be to build strong social movements united by a broad vision for Indigenous sovereignty, economic democracy, and decarbonization.

As well, a few specific strategies can more directly push the NDP (and possibly other parties) to be an aid rather than an obstacle to a just transition. First, we can organize fellow workers within and outside of unions. Rather than act as a conservative, don't-rock-the-boat influence on the NDP and its politics, a revitalized labour movement could instead push for the kinds of pressing changes that are needed.

Second, carefully considered electoral campaigns—where the candidates are driven by, accountable to, and in relationship with local social movement organizations and Indigenous Nations—can be helpful, even aside from their aim of taking elected office. These campaigns can be a strong vehicle for engaging and training new organizers who then stay involved in social movement efforts long after.[*] Elections also provide a useful opportunity for talking about many issues and their interconnections at once, which can begin to shift expectations about what kinds of solutions are posed. These conversations could, over time, help create better conditions and more transformative politics within and outside the NDP and other parties.

We should not rule out the possibility of building an entirely new

---

[*] From 2016 to 2020, the Bernie Sanders campaign helped facilitate past campaign volunteers to support actions like labour strikes and to build new local political organizations working outside the electoral sphere. This success has been replicated in smaller ways with runs for office in Canada by, among others, Matthew Hamilton, Paige Gorsak, Anjali Appadurai, and Paul Taylor.

political party. A party willing to take positions truly in line with a decolonial just transition would create a threat to the NDP's voter base from the left, and serve as a powerful prod for the NDP to stake out more radical positions.[†] A party-building effort could also serve as an important training ground for new activists, and add to the broader ecosystem of organizations and networks needed to win a just transition.

Finally, we can also exert pressure on the NDP externally by organizing to shape the party's actions at key moments, like conventions or leader elections. The Sunrise Movement in the United States, for example, has forced the Democrats to promise much bolder climate policy and adopt some pieces of a visionary Green New Deal. The movement's campaigning has featured creative mass actions linked to critical decision points and massive volunteer power to support challengers to key conservative incumbent candidates. Much more will be needed to shift the Democratic party enough to open up transformative change. But the combined efforts of the Sunrise Movement and other social movement organizations like the Justice Democrats and Democratic Socialists of America have started to act as a counterweight to the massive influence of Wall Street on the party.

We have focused on the NDP here, since it is the most likely existing party to serve as a useful force towards a decolonial just transition. The Green Party is both much smaller and more ideologically varied than the NDP, with a higher portion of its leaders and members supporting staunchly colonial and capitalist positions unaligned with a decolonial just transition. Many of the same strategies we have named for the NDP could be applied to the Greens, but we estimate a higher amount of effort would be needed relative to the potential payoff. However, an avowed ecosocialist, Dimitri Lascaris, was almost elected leader of the federal Greens in 2020, indicating that a base of support is present in the party's membership.

We do not see much evidence that the Liberal and Conservative parties can be pushed to become aligned to our vision, but it can be useful

---

† It is useful to draw on the example given by the opposite end of the spectrum. Federally the Conservative party has seen a rival party emerge to contest its votes, the People's Party of Canada, led by Maxime Bernier. Both the Conservative Party and the People's Party have terrible policies and terrible histories, so please don't mistake this example as any kind of endorsement. However, the effect of having the PPC contest races and erode the Conservative party's voter base means that the Conservative party now has to consider the PPC policies and support when it crafts policy and stakes out positions.

to assess opportunities to reduce the harm they pose while in power and win concessions that can open up further space to build power. Minority governments have in the past been pressured to implement substantial reforms—medicare in the 1960s and billions for housing and transit in 2005 were both achieved under minority Liberal governments.[8] To win harm reduction or useful concessions, we can look for points of division within these parties and organize behind "transformative and winnable" policies that social movement organizations and smaller opposition parties can align on.

## We Need Everyone

In the simplest terms, what is needed above all to build a decolonial, liberatory, and climate-safe world is for more people to work together towards one. If you take away a single thing from this book, it is this: try to find a way to become engaged in shared work towards this future if you are not already. By nature, it is not easy to win collective struggles. It requires challenging people and organizations with much more consolidated power. After years and decades of being involved in lots of different parts of this work, we still find ourselves learning and humbled daily by the scale of this challenge. We are far from knowing all of the answers, but here are some of the lessons we wish we'd known about this work when we first got involved.

**1. You can't struggle alone.** Syed Hussan, a long-time community organizer who directs the Migrant Workers Alliance for Change, explained this principle beautifully in September 2016, as many people in Toronto were being newly politicized around racial justice and other issues:

> I see so many new faces, bright, powerful, fierce minds at public actions and I know that many of you aren't in organizations. I am writing today, as someone who's been around barely a minute longer than you, to say that you must.
>
> The struggle is collective. We need others to inspire us, challenge us and change us. Join existing organizations or collectives, change them if you hate them, or start new ones. Organizing is a skill set, it's not just a set of ideals, and those skills must be honed. There are no schools, and few mentors.
>
> In my experience, individuals in organizations, affinity groups,

and collectives remain politicized for longer, because we need a counter-balance to the rest of society's imposition of a very different truth. Part of what neoliberal capitalism and colonialism have done is individuated us. At best, we may have our nuclear family's support, or have a partner, but we are told that we have to always look out for #1. That each of us needs to make decisions that improve our individual life. At the same time, we are told, no one person can change anything, and so we find ourselves in a bind.

.Here's the thing, one person can't change anything. But a few people, working together, in comradeship certainly can. If there is a hope in hell of us transforming our society, and building the kinds of worlds we want to live in, we need masses of people organized, disciplined and militant.[9]

We are thinking of organizations very broadly here—any group of people acting strategically in line with a decolonial just transition. Joining up does not necessarily mean upending your life to spend endless hours helping to run an activist group. That's not feasible for many of us, alongside fighting to make rent and buy groceries, and it may also not be where you can have the most impact. Individuals' contributions to collective power can come in many forms. Often our strongest levers are in the communities, workplaces, and social networks we are already in day to day.

**2. There are a thousand ways to "get involved"; consider what roles you might be best positioned to play.** Social movements are often portrayed in a misleading way. The narratives of Western individualism generally overemphasize a few figureheads and leaders, and social movements wins are recast as the miracle work of extraordinary superhumans. Martin Luther King Jr., Greta Thunberg, and Nelson Mandela are all amazing people who changed the course of history, but retellings that focus on them usually miss the broader ecosystem that make social movement wins possible. Behind these figureheads, there are thousands of people going to meetings, canvassing the public, organizing food and housing for participants, doing research, and playing hundreds of other roles.

So when we recommend joining an organization, that can mean an activist group, a community monitoring program in your First Nation, your workplace union, a student group, a mutual aid collective, a tenants' union, an Indigenous language school, a volunteer chapter of a not-for-profit, or a media collective. But it can also mean finding a friend or two to plan a fundraiser for other groups, to work on getting a solar project or food garden built in your Nation, or to talk to others about a problem in your neighbourhood or place of work.

Angele Alook notes that this kind of collective work is something Indigenous youth often learn growing up in First Nations reserve communities like her own (or as part of an urban Indigenous community, where she is raising her family). Even from a young age, Angele remembers helping out in the kitchen with the aunties when there was a funeral or gathering, knowing everyone in a collective has a role to play and everyone's role is important:

> I recall watching my children work in the garden with their grandparents or watching my nephews and nieces prepare a moose to share out the food with family. Children grow up learning to share resources with relations. I can recall organizing around trauma and loss so many times—and in opposition to colonial violence, and it's time we start organizing around life and resurgence. Contributing to communities of care from a young age prepares many Indigenous peoples to one day be leaders in economies of care—it is these communities of care and economies of care that lead to social movements. This may be the reason Indigenous youth are at the forefront of climate justice movements around the world.

With all of the work that is needed, it can be hard to figure out where to focus your efforts. Aside from thinking about your skills and interests, consider what relationships and power structures you already encounter day to day; these are almost always where you have the most power to exert. Maybe you live near fossil fuel production, a large prison, or another site that is a core to the death economy—you could find others in your community who are interested in figuring out how to build an alternative in its place. (Aamjiwnaang and Sarnia Against Pipelines, for example, organizes in Ontario's "Chemical Valley," where 40 percent of Canada's petrochemical industry is located!) Maybe you are a precarious worker who sees your co-workers or peers struggling with similar issues; you could look for ways to counter them together. (The Brampton-based

Naujawan Support Network unites migrant workers to work together to recoup stolen wages from employers!) Maybe you are a parent whose child has in-school police officers, and you could support or start efforts to defund this program. (Parents in Saskatoon supported this work, led by Black students!)

Part of the reason it can be difficult to find these clear inroads to collective work is that we do not yet have enough organizational "infrastructure" in place to facilitate it. For example, suburban communities—especially suburban immigrant communities—have been all but ignored by more established organizations on the left in recent years, while conservative parties in Canada have put increasing resources into their multilingual media and canvassing.[10] Still, active and inspiring organizing is happening in these suburbs, in line with a just transition—but often in isolation without resources from other parts of social movements. We have already identified revitalizing unions as one way of rebuilding these inroads, and we will suggest more below. You may see one of these gaps in your community—and we encourage you to think about what collective work could be started or built on to fill it.

**3. This work is prefigurative—which means practising the good relations we need a decolonial just transition to be rooted in.** Building a healthy movement culture is ultimately also vital prefigurative work: we want the ways we work now to reflect, in miniature, the kind of relations in the world we are trying to create. Building the organizations and networks we need right now will involve sharing power, practising collective healing, and navigating conflict. These are all skills we will need on an even larger scale if we are going to live in a good way in balance with each other and the earth. It all comes back to what it means to meaningfully uphold Treaty and other agreements made with Indigenous peoples. As *The Red Deal* reminds us: "Unlike the European Westphalian state model of sovereignty—defined by exclusivity over territory—Indigenous treaties such as Gdoo-nagganinaa (which defined the shared territory as 'our dish') allow for diverse, overlapping Indigenous jurisdictions and sovereignties."[11] And as nehiyaw scholar Emily Riddle argues, "European political traditions would have us believe that being sovereign means asserting exclusive control over a territory," whereas Indigenous political traditions "teach us that it is through our relationship with others that we are sovereign, that sharing is not a sign of weakness but of ultimate strength and diplomacy."[12] We need to build movements and visionary alternatives in a way that upholds the original instructions of Treaties like Gdoo-nagganinaa, basing our decisions on moral responsibility and peaceful and mutual relations.

The Movement for Black Lives, an abolitionist coalition of more than fifty Black-led organizations in the United States founded in 2014, did years of groundwork that made the mobilizations in the wake of the police murder of George Floyd in May 2020 much stronger and more clearly defined. As the movement's fundraising coordinator Charles Long said that July: "We were poised to do this because we've been building trust together, learning how to be a movement of people who could govern ourselves as well as money, and make decisions around how to make our vision a reality."[13]

Many long-time community organizers have been raising the alarm that a failure to put relationships first has instead often led to our movements being unable to have generative conflict. Yotam Marom, an organizer and facilitator who has played key roles at Occupy Wall Street and in the Palestine solidarity group IfNotNow, argued in an article on this topic that "learning to have generative conflict isn't just a matter of healthy organizations, it a prerequisite to developing good strategy."[14] He has found that without building up the skills and relationships to do so, groups either generate "laundry lists" instead of focused strategies or they collapse at the first sign of differences. In an interview with us, Maya Menezes, an organizer who has worked with No One Is Illegal Toronto and The Leap, notes that at an individual level, "the fear of upsetting others by getting something wrong often means that people only make the safest decisions or they withdraw from this work altogether, because they have seen others ripping each other to shreds in public over small disagreements."[15]

As organizer and political strategist Ejeris Dixon notes: "Without shifting our focus to repairing our relationships, our movements will rot from the inside out."[16] The most important way to make this shift is to remember that we can only work at the speed of trust. It can be very easy to zero in on how much work is to be done and focus on the most "efficient" way to do it—and it can often seem like doing it on your own is the fastest. But if you are not helping others learn along the way, making space for everyone to make mistakes, practising working through conflict, and carving out space to share joy and grief and every emotion in between, you will get much less done in the long term.

**4. Seek out and help others find opportunities for training, mentorship, and developing strategy.** Social movement work is hard, and as Syed Hussan pointed out, it has "no schools, and few mentors." We need to build those schools and raise more mentors to fill the gap, but in the meantime, it is worth seeking out ways to learn more, whether by finding trainings, organizing convergences, or reading books. No small part of this puzzle is also building relationships between generations, between movements, and between kinds of movement work to share lessons and help unearth winning strategies. As many young people get involved in collective work for the first time, let's open up opportunities for our elders—particularly from the bigger and more powerful social movements that were built in the 1960s and 1970s—to pass on lessons.

**5. Solidarity is a verb—"our liberation is bound up together."**[17] The individualism that undergirds our society causes racism, homophobia, transmisogyny, and other prejudices to be too often understood as individual rather than structural. For white settlers especially, this needs deep attention. Speaking in 2015 to a chapter of Showing Up for Racial Justice, a group that organizes white people to take anti-racist action, artist and organizer Ricardo Levin Morales summed it up well:

> White people are taught that racism is a personal attribute, an attitude, maybe a set of habits. Anti-racist whites invest too much energy worrying about getting it right; about not slipping up and revealing their racial socialization; about saying the right things and knowing when to say nothing. It's not about that. It's about putting your shoulder to the wheel of history; about undermining the structural supports of a system of control that grinds us under, that keeps us divided even against ourselves and that extracts wealth, power and life from our communities like an oil company sucks it from the earth.
>
> The names of the euro-descended anti-racist warriors we remember . . . are not those of people who did it right. They are of people who never gave up. They kept their eyes on the prize—not on their anti-racism grade point average.

This will also be the measure of your work. Be there. No one knows how to raise a child but we do it anyway. We don't get it right. The essential thing is that we don't give up and walk away. Don't get me wrong. It is important to learn and improve and become wise in the ways of struggle—or of parenting. But that comes with time. It comes after the idea of not being in the struggle no longer seems like an option. . . .

You may not get the validation you hunger for. Stepping outside of the smoke and mirrors of racial privilege is hard, but so is living within the electrified fences of racial oppression—and no one gets cookies for that. The thing is that when you help put out a fire, the people whose home was in flames may be too upset to thank and praise you—especially when you look a lot like the folks who set the fire. That's OK. This is about something so much bigger than that.

There are things in life we don't get to do right. But we do get to do them.[18]

But just because white settlers are where the largest "deficit" in solidarity lies does not mean white people acting in solidarity with BIPOC-led movements is the most powerful or needed form. We need new, stronger interconnections between all struggles, and solidarity between communities of colour—beyond and within Canada's borders—is particularly powerful. For example, solidarity across Canada to support the farmers' protests in India in 2020 and 2021 helped secure their wins and formed new kernels that are sprouting into new organizing efforts to address issues faced by the Indian diaspora working-class here.[19]

## Movements Are Infrastructure!

One barrier to working to solve Paulo Freire's question—"What can we do today, so that tomorrow we can do what we are unable to do today?"—is that we are facing overlapping crises and so much is needed *today*. People often get involved in social movement work precisely because they feel this urgency—as did all of the author team writing this. Climate scientists tell us we have until 2030 to cut global emissions in half, Indigenous elders holding unique language and cultural knowledge have so much to share in limited time, people are facing growing houselessness and hunger now. Building sustainable, healthy, and scalable movement infrastructure doesn't mean stopping direct work on these urgent problems, but it does mean making decisions and looking for resources with an eye on where we will want to be in five or more years.

Wet'suwet'en-led blockades and the Quebec climate strikes were both able to make the strides they did because of years of deeply strategic (and often invisible) work to build movement infrastructure that provided pathways for many new people to get, and stay, involved. Here's a breakdown of some of the varied tasks that we see as contributing to building this infrastructure.

**Growing an ecosystem of organizations:** A variety of organizations and organizational structures need to be stewarded to play different roles. This will mean strengthening some existing organizations, reforming others, and building new ones altogether. A few of the largest gaps are as follows. First, we need more organizations that can meet people where they are by providing easy, consistent ways for new folks to get involved in a manner that is meaningful to them. This means facilitating specific blocs of people to learn and take action together, such as union members, residents in a neighbourhood, scientists, youth in a First Nations community, or members of a faith group. Second, organizations are needed to provide deeper movement-oriented training and education, like Indigenous language and culture schools, or cohort programs to share skills and build relationships across different parts of the movement.* Both of these gaps can be addressed through large, established institutions like unions, grassroots organizations, distributed networks with chapters in different communities, or hybrid models with paid staff accountable to a broader community.†

---

\* Relative to the United States, there are very few well-resourced paths in Canada to get a political education and develop skills for long-term social movement work. The institutions offering training in any large-scale way—that have built and stewarded programs, curriculum, organizations, and capacity for this—are limited to a few small programs like Next Up or Tools for Change, a few programs within student unions and labour unions, and to some specific Indigenous and diaspora community organizations.

† Useful examples to emulate include the Toronto Workers' Action Centre Fight for 15 and Fairness campaigning, Lancaster Stands Up, and Indigenous People's Power Project (IP3). As one example, Lancaster Stands Up has a couple of paid staff to act as volunteer coordinators who can be consistent in following up with and getting to know new volunteers in order to support them and match them with work that suits their interests and schedule. This has helped the group avoid burnout among core organizers and grow to hold dozens of volunteer organizers across multiple working groups and chapters.

**Learning to build coalitions in a good way:** An essential part of any ecosystem is the connections between components. This is true for social movements too, and ours in so-called Canada have been too siloed despite obvious common interests. As Menezes put it to us in an interview when describing her efforts to build new cross-movement coalitions: "What has been keeping me up at night lately is the worry that people don't understand that we're all talking about the same thing." And indeed, cross-movement coalitions are behind almost every movement success story. Building more of them—at a variety of scales, forms, and levels of coordination—will help us forge a political alignment that can counter the current ruling colonial and capitalist alignment.

In the 2012 Quebec student strike, a little-known factor critical to its success was that for the three years prior, the Red Hand Coalition of 125 unions, community organizations, and student groups against the Quebec government's austerity measures had formed and begun deliberating on how to prioritize different strategies. The coalition was able to agree on the priority of stopping the tuition hike and putting forward a vision for free tuition. This didn't mean everyone dropped everything to support the goal, but it did mean groups carved out resources ahead of time to support the student strikes. We need to dedicate resources, time, and considerable practice to constructing coalitions that can help us pull in the same direction, like the Red Hand did.

Menezes argues this is simpler than it can seem; it means building relationships. "This means getting better at speaking together to map the common ground in our goals and long-term visions. It also means learning to move through conflict as a generative exercise that sharpens our political analysis and our tactics, instead of as an interpersonal barrier to connection and trust. And instead of outlining never-ending lists of shared public declarations as coalitions tend to, the most effective path is often making clear plans that define how we will work together."

**Establishing physical spaces:** Shared gathering places facilitate stronger relationships and often make it easier for new people to get involved. These can include office spaces, community centres, dedicated meeting spaces, temporary protest and blockade camps, and even shared living spaces. They allow for meetings and events to more easily happen, for meals to be shared, for services to be provided that build bridges with local communities, and for different groups to meet each other. Workers'

Action Centres, Native Friendship Centres, movement houses for organizers, and land-based camps have played important roles in facilitating
movements. These models can all be scaled up and shared. For example, the Halifax Workers' Action Centre was founded in the fall of 2017
by Solidarity Halifax and the Halifax-Dartmouth & District Labour
Council (a coalition of unions), and was directly inspired and facilitated
by the Toronto Workers' Action Centre, which assisted in setting the
Halifax centre up.[20]

**Building communications, media, and research infrastructure:** Few
media outlets or communications channels in Canada can give the ideas
behind a decolonial just transition a fair shake. The "legacy" media in
Canada aren't, in general, working in the interests of collective efforts
for decolonization and decarbonization. In fact, they more often than
not take positions and publish coverage in the interests of colonialism
and big business. The decolonial left needs to shift existing media and,
crucially, build its own media organizations. Something that gives us tremendous hope is that this shift has been happening, with an explosion
of new decolonial left media organizations in recent years and renewed
popularity of older outlets with similar politics.

These media workers have been chipping away at the colonial, neoliberal consensus and showing different possibilities. From *One Dish
One Mic* radio and the *Media Indigena* podcast talking decolonial politics, to the podcasts in the socialist Harbinger Media Network. Print
publications that carefully consider movement strategy like *Briarpatch,
Canadian Dimension, Upping the Anti,* and *Midnight Sun*. Militant labour
journalism like *Rank & File*, to investigative climate reporting at the
*Narwhal*, to left YouTubers and video makers like the folks at the *Breach*.
Abundant media is pushing back against the dominance of Canadian
fossil capitalism. As much as possible, movement and independent
media alike should tap into a decolonial left populist frame—drawing
parallels across experiences and now-dispersed issue-based movements
in order to present a shared narrative of "us"—workers, Indigenous peoples, migrants and displaced folks, women, LGBTQ2S+, disabled, and
young people—and "them"—a small group of settler capitalists. Using
populism also means communicating problems in accessible and clear
ways that identify both simple solutions and ways to get involved more
deeply.

There is also a great need for organizations that do movement-supporting research, formulating detailed policy proposals and both quantitative and qualitative analysis to inform movement strategy. We especially need new capacity for imagination and ambition in proposing a world significantly different and much better than the one we currently inhabit. Marxist literary theorist Fredric Jameson famously mused that "it's easier to imagine an end to the world than an end to capitalism." This capitalist realism—the idea that things as they are might be terrible, but nothing beyond it can be conceived of or entertained—has had awful effects on our collective ability to imagine a world beyond capitalism, colonization, and climate collapse.[21] But Jeremy Corbyn's left populist project in the United Kingdom was supported by the think tank Common Wealth—a heartening example of the kinds of alternatives research we could be putting forward here.[22] We need networks of think tanks like this putting their policy proposals in dialogue with each other, debating their merits, and making public arguments about their various trade-offs and benefits.

For this research to be effectively formulated and useful for front-line communities, movement groups, unions, and Indigenous Nations, it needs to be done in relationship with these groups. At present, only a few such organizations and academic clusters exist. The Canadian Centre for Policy Alternatives, the Yellowhead Institute, Parkland Institute, Broadbent Institute, Indigenous Climate Action, and the Canadian Society for Ecological Economics are some think tanks engaged in formulating proposals and advocating for policy that puts people and the planet first. And groups like Haíɫzaqv Climate Action conduct community-led research to inform policies, reassert their own laws, and pilot projects tailored to their Nations.

## Liberating Time, Land, and Money

Building up movement infrastructure and organizational support costs money, requires many hours of unpaid labour that most people can't currently squeeze in alongside their precarious livelihoods, and demands physical space and land. The powerful people's movements in Canada in the 1960s and 1970s benefited from strong supporting organizations, like unions or the National Action Committee on the Status of Women, whose resources have since been greatly eroded under neoliberalism. For all the conspiracies of the right about "foreign-funded environmentalist plots" to stop the tarsands, the resources of climate justice and Indigenous rights focused organizations in Canada pale in comparison to what is available for other issues as well as what is available for similar work in other countries.[23] Meanwhile, the companies and individuals who stand to profit from continued extractive colonialism in Canada have decades of profit hoarding at their disposal to defend their interests. Considering how unequal the playing field is, community-based initiatives and organizing are winning some amazing things. Strategically redistributing available resources to help these efforts grow provides one of our best shots at winning.

Beyond building more movement infrastructure, redistributing resources is also key to addressing *who* can spare time and mobilize resources to organize and build community-based alternative institutions. This ability is unfortunately often warped along class, racial, and other lines—where relatively wealthy and disproportionately white individuals are often able to spend the most time organizing. In order to uphold the principle of "Nothing about us without us!" it is critical to ensure Indigenous Nations and other groups most affected by colonialism and capitalism lead the work for a just transition. We need to redistribute resources to make it more sustainable for communities on the front lines of the issues to lead. Winning stronger workers' rights and a four-day workweek (as discussed in chapter six) is one way to help guard people's time to fight for our collective interests. Here are a few more.

There is a form of spiritual bankruptcy that comes with hoarding wealth. In other words, as the impacts and linkages between climate change, capitalism, systemic racism, and colonialism have been brought to the surface of political discourse, it has become a lot more difficult for people with excess wealth to avoid confronting the fact that it is immoral and socially destructive to keep it. This has opened the possibility of organizing people with wealth and class privilege to learn about inequality and redistribute their wealth. A US organization with this goal called Resource Generation has mobilized its young members to redirect their inherited wealth or other money they can access towards "poor and working-class led economic and racial justice movements." Tens of millions of dollars have been redistributed since 2014 in support of Movement for Black Lives organizations. Resource Generation's work includes providing training to members, helping them encourage family members and friends to follow suit.

One of our co-authors, David Gray-Donald, along with a small number of other interested folks, started a Canada-based version of this model called Resource Movement. Resource Movement looks to funds led by poor and working-class volunteer community representatives to make decisions about what projects to support, such as Groundswell Community Justice Trust Fund in Ontario. "A small group of wealthy and class-privileged young people aren't going to dream up a new version of the economy," David told *Maclean's* magazine.[24] "We're going to take leadership from other people," specifically working-class communities most affected by various oppressions. Wealth is relative, but this practice of redistribution can be helpful even if you are far from the 1 percent. As the Groundswell fund puts it, "For those with extra to spend, it is important to consider how much is 'enough,' and what could be done if that money was pooled to help build organizations that could actually improve people's lives."[25]

A corollary of Resource Movement, a potentially even more transformative one, is the Treaty Land Sharing Network. This network was started by another of our co-authors, Emily Eaton, along with six other people, to connect farmers and other landholders with Indigenous peoples in Treaty 4 and Treaty 6 (spanning parts of Alberta, Saskatchewan, and Manitoba) who need safe access to land to gather plants and medicines, hunt, practise ceremony, and so on. As Angela Roque of the Anishinabek Nation Treaty Authority states:

> The Anishinabe require access to land in order to exercise their Treaty rights and meet the needs of the communities. The Treaty Land Sharing Network has not only opened access to privately held land, it has opened a possibility to build respectful and positive relationships based on the Treaty principles of mutual respect and mutual benefit.[26]

Another set of resources to shift is those held in social institutions like labour unions, not-for-profit organizations, and universities. These groups have immense organizing and mobilizing power, and despite the impacts of austerity, relatively large budgets. Worker committees, community partners, resolutions, and new candidates for the executive can all help push labour unions to use resources in support of movements for a just transition. This can take the form of the Bargaining for the Common Good strategies or drives to organize precarious workers from chapter six, but also directly using labour union funds to support a broader ecosystem of movement organizations. At universities, academics, workers, and students can strategically siphon class time, in-kind resources, and research to support movement work. For example, the educational programming offered by Unist'ot'en Camp was supported in part by a course created in collaboration with the University of Northern British Columbia. Not-for-profits can similarly be pushed to dedicate resources and pursue strategies aligned with a decolonial just transition by their workers or community partners.[27]

## Looking Back to Go Forward

The crises we are facing can easily be debilitating. It is hard to imagine gathering the collective political power needed to build a new future that does not keep stealing land and life. In our own experiences, collective work towards a better future is the most powerful antidote for these feelings. This work by nature builds relationships and provides meaning in a time when our economic systems increasingly push us to be isolated and nihilistic. It gives us a community where we can work through the grief of this time together, and find joy in actively imagining and putting into practice pieces of a better world.

Climate change is far from the first world-ending threat human populations, particularly communities of colour, have faced. It can be helpful to hold this fact close. There is a long history of people in precarious, unjust, and seemingly impossible situations not knowing what will happen, but struggling and organizing to build a future anyway.[28] Indigenous cultures, languages, and peoples are alive and vibrant in spite of centuries of colonial efforts to wipe them out.

And in this history of struggle, there are so many important lessons on how we can go forward. We have found it grounding to see yourselves as part of, and accountable to, a long history of social movement work and prefigurative alternatives. A lot of the impacts of climate change and colonialism have already come to pass, but many are reversible and the worst-case scenarios are far from foregone conclusions.

These long histories of resistance also teach us that change is unlikely to be linear. We can imagine (as we did in chapter six) using the seeds of struggles for a just transition we already have today to get much closer to our vision and to build social movement infrastructure in the next few years. If we are successful in even some of our efforts to erode the power of fossil fuel companies, forge unlikely coalitions in front-line fossil fuel communities to fight for new paths forward, stand together to demand Indigenous peoples' land rights be fully upheld in more places, build strong unions willing to fight for systemic change, and free up billions of public dollars currently causing harm, many of the policies and projects we've proposed in this book will be within reach. Social movement momentum can take on exponential properties, and so even larger shifts to a fully decolonial future outside of capitalism may open up as well. What we know for sure is that we won't win a decolonial just transition that uplifts life if we are timid. We must be bold. Don't be afraid to dream big and loudly demand this livable and just future.

Finally, these histories teach us that if we get there, it will be because we have put living in right relation to each other and the natural world and returning to the sacred promises we made in the Treaties at the root of our struggle. It's on days when we practise these good relations, or see others do the same, that a decolonial and climate-safe future, where a good life is available to all, feels easiest to imagine.

# ACKNOWLEDGEMENTS

Much of the work of building climate justice in so-called Canada will come down to the challenge of forging deep collaboration across different communities and geographies, and in a lot of ways writing this book was a tiny microcosm of this. We are a diverse team of people who didn't know each other very well before this project began and through hundreds of hours in remote meetings and shared documents during a global pandemic we shared stories and big feelings, learned to talk through hard questions, and leaned on each other's varied experiences and skills. We came out the other end as dear friends. We hope that this book helps people imagine and shape a better world that is similarly relational, where we can take care of one another and the earth.

We would like to thank Derek Nepinak, Vickie Wetchie, and Maya Menezes for generously sharing your wisdom and experience in interviews with us—we learned a lot from you and hope your insights are adequately reflected in the book. We acknowledge the Corporate Mapping Project (CMP) for funding the research for this book. The CMP is a research and public engagement initiative investigating the power of the fossil fuel industry in Western Canada and is jointly led by the University of Victoria, the Canadian Centre for Policy Alternatives, and the Parkland Institute.

Between the Lines press has been a joy to work with, and we are grateful to them for taking on this book and for the work of their skilled staff: Amanda Crocker for helpful early big-picture direction, Devin Clancy for a fabulous book design, and Karina Palmitesta for getting our book out to booksellers. Special thanks to Tilman Lewis for his wise guidance and careful work editing and copy editing the manuscript. We are honoured to have the beautiful art of Chief Lady Bird on the book's cover. Several people read drafts of chapters and offered their advice and feedback, including James Wilt, Karey Brooks, Avi Lewis, Jacob Crane, and Ron Lameman, who also helped us with nehiyawewin (the Cree language). We thank them for their input but take full responsibility for any remaining errors, omissions, or oversights.

**Angele Alook:** I want to thank York University for supporting this book, especially for offering me special research access to my office so I always had someplace quiet to write during the pandemic lockdowns. I feel blessed to have been raised in Bigstone Cree territory by my parents Donald and Bertha Alook; your prayers and commitment to community is why I do this work. I am most grateful to nohkom Mary Young for taking me berry picking and always feeding me from the land as a child. I give thanks for having a family that speaks nehiyawewin. Jamie Taylor and Soleil Taylor need special thanks for all their care work, which gave me time to write. To Solomon Taylor, my youngest child, thank you for bringing home sage in your pockets and smudging the family; you are the reason we must protect Mother Earth.

**Emily Eaton:** I am grateful to have a tenured position at the University of Regina within an academic environment that values community-engaged scholarship and service. Relationships of love, struggle, and solidarity (mum, dad, Oliver, Val, Nick, Andrew, Karen, Marc, Jesse, Trish, Don, Mary, Amy, Martha, Naomi, Hillary, Autumn, and Shaid, to name a few) give me joy and keep me going while the world tumbles forward in increasingly painful and chaotic ways. This book is for our children. Simon Enoch and Violet Eaton-Enoch deserve special mention for accompanying me through the highs and the lows of the day to day.

**David Gray-Donald:** Deepest thanks to my friends in Regina, Montreal, Toronto, Winnipeg, and around so-called Canada who have taught me so much, shown such bravery, gotten into beautiful trouble, and inspired me to write. Thanks especially to Cha-nese Ila, for her companionship and encouragement while I was working on this book and as we navigated life and the pandemic. Thanks to my mum, Joy, for her support, and to my brothers. I dedicate this to my nieces.

**Joël Laforest:** I am immensely grateful to my friends and podcasting comrades who've made working, researching, and organizing in Calgary possible. Aaron, Brendan, Clinton, Doug, Elaine, Jessica, Josh, Karen, Kate, Levi, Patrick, Pat, Roberta, Sean, Stephen, Trevor, Tyler, André, William, and Will: you are the very best. Caillie Mutterback deserves special mention for her support, encouragement, and companionship over these difficult pandemic years.

**Crystal Lameman:** For nicawâsimisak (my children), who are my biggest sources of inspiration and who remain steadfast in their patience with me as I move to break generational curses. They remind me that this generational healing and justice work is for all that gives and breathes life, and for our ancestors who came before us, and for the generations yet to come. For nohtâwiypan (my late father), my biggest cheerleader, who started this journey with me, but who has since returned to the stars and the universe that never goes dark. For okâwîmâwaskiy (Mother Earth), my namesake, we do this for you. ninanâskomowak (I am thankful and grateful) to amiskosâkahikan nêhiyaw peyakôskân, ostêsimâwoyasi-wêwin nikotwâsik (Beaver Lake Cree Nation, Treaty No. 6), ekwa nîkânîwin (and my leadership), for supporting me in this endeavour. Finally, gratitude to my mentors, colleagues, allies, supporters, and those who love me.

**Bronwen Tucker:** I want to thank my friends and mentors for teaching me in so many ways that we are powerful if we struggle together and that we can stand on the shoulders of many others who did the same in the face of past world-ending villains. I'm also so grateful for eleven years of beloved roommates who have welcomed, held, gossiped, and shared with me and especially Tim, Cole, Del, Emma, and Carter for the last very wild three. Thank you to my family for everything.

# GLOSSARY OF CREE TERMS

| | |
|---|---|
| askiy oma | this land |
| atoskewin | to work, or to do work |
| kitimâkis | they are poor |
| kitimâkisowin | being poor or unhappy |
| manâcihtâwin | respect |
| miyo-ohpikinâwasowin | good child-rearing |
| miyo-pimatisiwin | living a good life |
| miyo-wîcihtowin (miyo-wîcêhtowin) | good relations and unity; the act of having good relationships |
| mosom | grandfather |
| nehiyaw (nêhiyaw) | Cree |
| nehiyawewin (nêhiyawêwin) | Cree language; Cree language speaker |
| nehiyaw-iskwewak | Cree women |
| nohkomak | my grandmothers |
| nohkômpan | my late grandmother |
| nohtâwiypan | my late father |
| pimâcihowin | livelihood |
| pimâtisiwin | life |
| sihtoskâtowin | coming together in mutual support; pulling together for survival |
| wânaskêwin | living in peace and harmony |

# NOTES

## Introduction

1  Tom Flanagan, *Resource Industries and Security Issues in Northern Alberta* (Calgary: Canadian Defence & Foreign Affairs Institute, 2009).

2  Marci McDonald, "The Man behind Stephen Harper," *Walrus*, October 2004, thewalrus.ca.

3  Ploy Achakulwisut and Peter Erickson, "Trends in Fossil Fuel Extraction: Implications for a Shared Effort to Align Global Fossil Fuel Production with Climate Limits," SEI Working Paper, 2021, doi: 10.51414/sei2021.001; Alex Hemingway, "One Year Later: Canadian Billionaire Wealth Up by $78 Billion," *Policy Note* (Canadian Centre for Policy Alternatives), April 14, 2021, policynote.ca.

4  Flanagan, *Resource Industries and Security Issues*, 4.

5  Pamela Palmater, "Harper's 10 Year War on First Nations," *Harper Decade*, July 16, 2015, theharperdecade.com.

6  William Carroll, ed., *Regime of Obstruction: How Corporate Power Blocks Energy Democracy* (Athabasca University Press, 2021), aupress.ca.

7  "History and Context," Climate Justice Alliance, climatejusticealliance.org.

8  "Guidelines for a Just Transition towards Environmentally Sustainable Economies and Societies for All," International Labour Organization, 2015, ilo.org.

9  For a good exploration of degrowth, see: Matthias Shmelzer, Andrea Vetter, and Aaron Vansintjan, *The Future Is Degrowth: A Guide to a World Beyond Capitalism* (Verso, 2022), and the website degrowth.info.

10  "Canada's Fair Share towards Limiting Global Warming to 1.5°C," Climate Action Network—Réseau Action Climat, 2019, climateactionnetwork.ca.

11  Angele Alook, Ian Hussey, and Nicole Hill, "Indigenous Gendered

Experiences of Work in an Oil Dependent, Rural Alberta Community," in *Regime of Obstruction*, ed. Carroll, 331–53.

12 Simon Evans, "Analysis: Which Countries Are Historically Responsible for Climate Change?," Carbon Brief, October 5, 2021, carbonbrief.org.

13 See: Harsha Walia, *Undoing Border Imperialism* (AK Press, 2013); Max Ajl, *A People's Green New Deal* (Pluto Press, 2021); and "Peoples Agreement," World People's Conference on Climate Change and the Rights of Mother Earth, Cochabamba, Bolivia, April 2010, peoplesagreement.org.

14 Russ Diabo, "Putting Our House in Order—A Truth before Reconciliation Publication," Media Co-op, September 11, 2020, 11, mediacoop.ca.

15 Jonathan Matthew Smucker, "Yes, Populism," blog, July 8, 2012, jonathansmucker.org.

## 1. No More Broken Promises

1 This section draws from: Crystal Lameman, "Sakaskiniy Miyomahciho Pihkohtawin: Iyinaysiynew Miyoaskewpimatisiwin; Realizing Holistic Wellness: Meaningful Indigenous Land Based Practices" (University of Alberta, M.Ed. Capping Paper, 2019).

2 "Carry the Kettle First Nation (Carry the Kettle) Case Backgrounder," Carry the Kettle First Nation, cegakin.com; *Yahey v. British Columbia*, 2021 BCSC 1287 (CanLII), canlii.org.

3 "B.C., Blueberry River First Nations Reach Agreement on Existing Permits, Restoration Funding, British Columbia," BC Gov News, October 7, 2021, news.gov.bc.ca.

4 Paige Parsons, "Appeal Court Reverses Rare Advance Cost Award to Alberta First Nation," *CBC News*, June 30, 2020, cbc.ca.

5 *British Columbia (Minister of Forests) v. Okanagan Indian Band*, 2003 SCC 71 (CanLII), [2003] 3 SCR 371, canlii.ca.

6 *Anderson v. Alberta*, 2022 SCC 6 (CanLII), canlii.ca.

7 Claudia Sobrevila, "The Role of Indigenous Peoples in Biodiversity Conservation: The Natural but Often Forgotten Partners," World Bank, May 2008, documents.worldbank.org; Rishabh Kumar Dhir, Umberto Cattaneo, Maria Victoria Cabrera Ormaza, Hernan Coronado, and Martin Oelz, *Implementing the ILO Indigenous and Tribal Peoples Convention No. 169: Towards an Inclusive, Sustainable and Just Future* (International Labour Organization, February 2020), ilo.org.

8 Graeme Reed et al., "Towards Indigenous Visions of Nature Based Solutions: An Exploration into Canadian Federal Climate Policy," Climate Policy 22, no. 4 (2022), 514–33, doi: 10.1080/14693062.2022.2047585.

9 Dallas Goldtooth, Alberto Saldamando, and Kyle Gracey, *Indigenous*

*Resistance against Carbon* (Oil Change International, August 2021), ienearth.org.

10  As recalled by Morris in his account of events, published 1880. The Hon. Alexander Morris, *The Treaties of Canada with the Indians of Manitoba and the North-West Territories, Including the Negotiations on which They Were Based, and Other Information Relating Thereto* (Toronto: Willing & Williamson, 1880), 211, canadiana.ca.

11  Heidi Stark, "Respect, Responsibility, and Renewal: The Foundations of Anishinaabe Treaty Making with the United States and Canada," *American Indian Culture & Research Journal* 34, no. 2 (2010): 147.

12  Gina Starblanket, "Crises of Relationship: The Role of Treaties in Contemporary Indigenous-Settler Relations," in *Visions of the Heart: Issues Involving Indigenous Peoples in Canada*, ed. Gina Starblanket, David Long, and Olive Patricia Dickason (Oxford University Press, 2019), 23.

13  Starblanket, "Crises of Relationship," 19.

14  *Report of the Royal Commission on Aboriginal Peoples*, 1996, bac-lac.gc.ca, Vol. 2, 43, cited in Starblanket, "Crises of Relationship," 20.

15  "Genocide and Indigenous Peoples in Canada," *Canadian Encyclopedia*, last modified January 18, 2021, thecanadianencyclopedia.ca.

16  Cliff Atleo, "Between a Rock and Hard Place: Canada's Carbon Economy and Indigenous Ambivalence," in *Regime of Obstruction*, ed. Carroll, 356.

17  Andrew Curley and Majerle Lister, "Already Existing Dystopias: Tribal Sovereignty, Extraction, and Decolonizing the Anthropocene," in *Handbook on the Changing Geographies of the State*, ed. Sami Moisio, Natalie Koch, Andrew E.G. Jonas, Christopher Lizotte, and Juho Luukkonen (Elgar Online, 2020), 260.

18  Bonita Lawrence, "Regulating Native Identity by Gender," in *Gender and Women's Studies: Critical Terrain Second Edition*, ed. Margaret Hobbs and Carla Rice (Toronto: Women's Press, 2018), 326.

19  Alex Wilson, "Skirting the Issues: Indigenous Myths, Misses, and Misogyny," in *Keetsahnak: Our Missing and Murdered Indigenous Sisters*, ed. Kim Anderson, Maria Campbell, and Christi Belcourt (Edmonton: University of Alberta Press, 2021).

20  For example, Leanne Betasamosake Simpson notes that Nishnaabeg/Ojibwe men and women both "hunted, trapped, fished, held leadership positions, and participated in warfare, as well as engaged in domestic affairs and looked after children, and they were encouraged to show a broad range of emotions and to express their gender and sexuality in a way that was true to their being, as a matter of *both principle and survival*." (Leanne Betasamosake Simpson, "Centring Resurgence: Taking on Colonial Gender Violence in Indigenous Nation Building," in *Keetsahnak*, ed. Anderson, Campbell, and Belcourt, 220.)

21 Audra Simpson, "The State Is a Man: Theresa Spence, Loretta Saunders and the Gender of Settler Sovereignty," *Theory & Event* 19, no. 4 (2016): 1.

22 Simpson, "The State Is a Man," 3.

23 National Inquiry into Missing and Murdered Indigenous Women and Girls, *Reclaiming Power and Place*, Executive Summary of the Final Report, mmiwg-ffada.ca.

24 Hayden King, Shiri Pasternak, and Riley Yesno, *Land Back: A Yellowhead Institute Red Paper* (Yellowhead Institute, 2019), 25, yellowheadinstitute.org.

25 Troy Stuart, Just Powers / Future Energy Systems: Intermedia & Documentary—iDoc project (University of Alberta). On location at Bigstone Cree Nation, Alberta, February 27, 2018.

26 Angele Alook, Sheila Block, and Grace-Edward Galabuzi, *A Disproportionate Burden: COVID-19 Labour Market Impacts on Indigenous and Racialized Workers in Canada* (Canadian Centre for Policy Alternatives, 2021), policyalternatives.ca.

27 Melina Laboucan-Massimo, "Lessons from Wesahkecahk," in *Whose Land Is It Anyway: A Manual for Decolonization*, ed. Peter McFarlane and Nicole Schabus (Federation of Post-Secondary Educators of BC, 2017), 37, fpse.ca.

28 Indigenous Climate Action conducted an in-depth critical analysis of the two policies known as the Pan-Canadian Framework on Clean Growth and Climate Change (2019) and A Healthy Environment, A Healthy Environment (2020) "to investigate whether they take aim at the root causes of climate change and both respect and meaningfully include Indigenous Peoples and our rights, knowledge and approaches to climate action" (6). *Decolonizing Climate Policy in Canada: Report from Phase One* (Indigenous Climate Action, 2021), indigenousclimateaction.com.

29 *Decolonizing Climate Policy in Canada*, Indigenous Climate Action, 6.

30 *Decolonizing Climate Policy in Canada*, Indigenous Climate Action, 5, 46.

31 King, Pasternak, and Yesno, *Land Back*, 25.

32 Sylvia McAdam (Saysewahum), *Nationhood Interrupted: Revitalizing nêhiyaw Legal Systems* (UBC Press / Purich Publishing Ltd., 2015), cited in King, Pasternak, and Yesno, "Land Back," 53.

33 Mike Gouldhawke, "Land as a Social Relationship," *Briarpatch: The Land Back Issue*, September–October 2020, 11–13.

34 "Indigenous Principles of Just Transition," Indigenous Environmental Network, no date, ienearth.org.

35 Derek Nepinak, online interview with authors, April 9, 2021.

## 2. Delay and Deny

1 Katrin MacPhee, "Canadian Working-Class Environmentalism, 1965–1985," *Labour / Le Travail* 74 (Fall 2014), 131–33, id.erudit.org; John J. Lee,

"We Warned about Pesticides, Chavez Says in L.A.," *Los Angeles Times*, March 31, 1989, latimes.com.

2  As recently as 2021, a bill titled "The Endless Frontier Act" was introduced to the United States Congress.

3  "Indian Act," *Canadian Encyclopedia*, last modified December 16, 2020, thecanadianencyclopedia.ca.

4  Harold Cardinal, *The Unjust Society*, 2nd ed. (Vancouver: Douglas & McIntyre, 1999 [1969]), 1.

5  Cardinal, *The Unjust Society*, 2.

6  Edward Allen, "Letter from British Columbia: Reflections on the 40th Anniversary of the Calder Decision," *Northern Public Affairs*, September 2013.

7  Atleo, "Between a Rock and a Hard Place," 364.

8  Mackenzie Valley Pipeline Inquiry, *Proceedings at Community Hearing* (Vancouver: Allwest Reporting), Fort Good Hope, NWT, August 5, 1975, Vol. 18, 1769–71, pwnhc.ca. Available to view: Canadian Broadcasting Corporation, "Dene Chief Frank T'Seleie—Mackenzie Valley Pipeline/ Gas Project in 1975," posted by YouTube user TheUploadwordpress, April 6, 2013, 2:36 to 3:28, youtu.be/pohp-gYL1Io.

9  Glen Sean Coulthard, *Red Skin, White Masks: Rejecting the Colonial Politics of Recognition* (University of Minnesota Press, 2014), 169. Emphasis in original.

10  Valerie Lannon and Jesse McLaren, *Indigenous Sovereignty and Socialism* (Toronto: Resistance Press, 2018), 59.

11  Russ Diabo, "From the 1969 White Paper on Indian Policy to Today's White Paper 2.0: The Federal Bureaucracy's 50 Year Plan," Truth Before Reconciliation Campaign presentation slides, May 2020, 4, mediacoop.ca.

12  Russ Diabo, "Platform: Truth Before Reconciliation," from Diabo's campaign for Assembly of First Nations Chief, 2018, russdiabo.com.

13  "What Is a Modern Treaty?," Land Claims Agreements Coalition, landclaimscoalition.ca.

14  Department of Justice and Inherent Right Directorate, *Guidelines for Federal Self-Government Negotiators (Number 1): Language for Recognizing the Inherent Right of Self-Government in Agreements and Treaties* (Department of Justice and Inherent Right Directorate, March 22, 1996), 1–13, mediacoop.ca.

15  West Coast Environmental Law, "Supreme Court of Canada Refuses to Hear Trans Mountain Appeal," news release, July 2, 2020, wcel.org.

16  "Delgamuukw Case," *Canadian Encyclopedia*, last modified January 11, 2019, thecanadianencyclopedia.ca.

17  Arthur Manuel, *The Reconciliation Manifesto: Recovering the Land,*

*Rebuilding the Economy* (Toronto, James Lorimer & Company, 2017), 215. Also cited in King, Pasternak, and Yesno, *Land Back,* 30.

18  Jaskiran Dhillon, Will Parrish, "Exclusive: Canada Police Prepared to Shoot Indigenous Activists, Documents Show," *Guardian,* December 20, 2019, theguardian.com.

19  This section is adapted from co-author David Gray-Donald's *Canadian Dimension* article "Climate Capitalism and 'Regimes of Obstruction,'" July 27, 2021, canadiandimension.com, a double book review of Donald Gutstein's *The Big Stall* (2018) and *Regime of Obstruction,* ed. William K. Carroll (2021).

20  For more, see: Tyler A. Shipley, *Canada in the World: Settler Capitalism and the Colonial Imagination* (Halifax & Winnipeg: Fernwood, 2020); Catherine Nolin and Grahame Russell, eds., *Testimonio: Canadian Mining in the Aftermath of Genocides in Guatemala* (Toronto: Between the Lines, 2021); Jeffery R. Webber and Todd Gordon, *Blood of Extraction: Canadian Imperialism in Latin America* (Halifax & Winnipeg: Fernwood, 2016); and many other publications.

21  Alejandro de la Garza, "Fossil Fuel Companies Are Still Influencing COP26, Despite Losing Their Official Role," *Time,* October 27, 2021, time.com; Global Witness, "Hundreds of Fossil Fuel Lobbyists Flooding COP26 Climate Talks," news release, November 8, 2021, globalwitness.org.

22  John-Henry Harter, "Environmental Justice for Whom? Class, New Social Movements, and the Environment: A Case Study of Greenpeace Canada, 1971–2000," *Labour / Le Travail* 54 (Fall 2004): 104, lltjournal.ca.

23  Carol Linnitt, "'Grassroots' Oil and Gas Advocacy Group Canada Action Received $100,000 from ARC Resources," *Narwhal,* June 24, 2020, thenarwhal.ca.

24  To understand more about the ideological function of industry-funded campaigns and philanthropy, see: Shane Gunster, Robert Neubauer, John Bermingham, and Alicia Massie, "'Our Oil': Extractive Populism in Canadian Social Media," in *Regime of Obstruction,* ed. Carroll; and Emily Eaton and Simon Enoch, "The Oil Industry Is Us: Hegemonic Community Economic Identity in Saskatchewan's Oil Patch," in *Regime of Obstruction.*

25  Shane Gunster, "Extractive Populism and the Future of Canada," *Monitor,* July 2, 2019, policyalternatives.ca.

26  Donald Gutstein, *The Big Stall: How Big Oil and Think Tanks Are Blocking Action on Climate Change in Canada* (James Lorimer & Company, 2018), 68.

27  Exxon owned 70 percent of Imperial Oil as of 2020. Murtaza Hussain, "Imperial Oil, Canada's Exxon Subsidiary, Ignored Its Own Climate Change Research for Decades, Archive Shows," *Intercept,* January 8, 2020, theintercept.com.

28   Seth Klein and Shannon Daub, "The New Climate Denialism: Time for
     an Intervention," Canadian Centre for Policy Alternatives, September 30,
     2016, corporatemapping.ca.

29   "Clean Growth: Building a Canadian Environmental Superpower,"
     Business Council of Canada, October 1, 2007, thebusinesscouncil.ca.

30   Jessica F. Green, "Does Carbon Pricing Reduce Emissions?: A Review
     of Ex-Post Analyses," *Environmental Research Letters* 16, no. 4 (2021),
     iopscience.iop.org.

31   An outlier to this trend is arguably Sweden, which as of 2020 had the high-
     est carbon taxes in the world, at US$126 / tonne $CO_2$ equivalent ($CO_2e$).
     Canada's price was about $22 in 2020. Emissions have been going down
     in Sweden, but while this is touted by some as a sign of success of carbon
     taxes, carbon taxes were one of several policy approaches the Swedish
     government has enacted. This makes it difficult to parse the exact effect
     of the carbon tax. ($CO_2e$ and $CO_2$ are sometimes used interchangeably
     in climate writing. Technically, $CO_2e$ includes carbon dioxide and other
     greenhouse gases, like methane, and compiles them all into one easy-
     to-understand number, whereas $CO_2$ only refers to that one gas, carbon
     dioxide, which is the main, but not the only, driver of global heating.)

32   Justin Trudeau, "Liberal Party of Canada Leader Justin Trudeau's Speech
     to the Calgary Petroleum Club," October 30, 2013, liberal.ca.

33   "Representing Justin Trudeau's new government were his top two polit-
     ical advisers, Gerry Butts and Katie Telford. With them was the woman
     who at the time headed Canada's public service, Privy Council Clerk
     Janice Charette. They were facing Brian Topp, a wily political tactician
     who recently stepped down as Alberta Premier Rachel Notley's chief of
     staff. At Topp's side were Richard Dicerni, then head of the Alberta public
     service, and University of Alberta environmental economist Andrew
     Leach, at the time chair of Notley's Climate Change Advisory Panel."
     Peter O'Neil, "The Inside Story of Kinder Morgan's Approval," *Vancouver
     Sun*, January 6, 2017, vancouversun.com.

34   Justin Trudeau, "Statement by the Prime Minister of Canada after
     Delivering a Speech to the Assembly of First Nations Special Chiefs
     Assembly," December 8, 2015, pm.gc.ca. Trudeau had mentioned the line
     about communities granting permission as early as 2013 (Jennifer Stahn,
     "Trudeau: 'Only Communities Can Grant Permission,'" *InfoTel News*, July
     24, 2013, infotel.ca) and continued into 2016 ("Trudeau: 'Governments
     Grant Permits, Communities Grant Permission,'" CBC, video clip, posted
     March 1, 2016, cbc.ca).

35   David Hughes, "Can Canada Expand Oil and Gas Production, Build
     Pipelines and Keep Its Climate Change Commitments?," Corporate
     Mapping Project, June 2, 2016, policyalternatives.ca. The Corporate

Mapping Project, investigating the power of the fossil fuel industry, is jointly led by the University of Victoria, Canadian Centre for Policy Alternatives, and the Parkland Institute.

36 "Oil Sands Operations: Bitumen Production," *Oil Sands Magazine*, last updated January 31, 2022, oilsandsmagazine.com.

37 Brandi Morin, "'It's Not Something That Happens Overnight': Alberta Premier Asks Indigenous Peoples to Be Patient While Government Works on Promises," *APTN National News*, September 15, 2015, aptnnews.ca.

38 Catharine Tunney, "Jim Carr Says Military Comments Not a Threat to Pipeline Protesters," CBC, December 2, 2016, cbc.ca.

39 Lee Wilson, "Saik'uz First Nation in B.C. Says No More Logging unless Companies Have Its Consent," APTN *National News*, October 19, 2021, aptnnews.ca. See also: Torrance Coste, "Double Standard: B.C. Requires Indigenous Consent for Forest Conservation but Not Logging," *Ricochet*, September 9, 2021, ricochet.media.

40 Wet'suwet'en Hereditary Chiefs, BC Civil Liberties Association, Union of BC Indian Chiefs, "Wet'suwet'en, BCCLA, and UBCIC Release Explosive Letter Revealing BC Solicitor General Authorizing RCMP Deployment, Contradicting Public Statements," news release, March 6, 2020, ubcic.bc.ca.

41 Russ Diabo, "Federal UNDRIP Bill C-15 Is an Attack on Indigenous Sovereignty and Self-Determination: Opinion," APTN *National News*, December 21, 2020, aptnnews.ca.

42 Marc Lee, *Dangerous Distractions: Canada's Carbon Emissions and the Pathway to Net Zero* (Canadian Centre for Policy Alternatives BC Office, June 17, 2021), corporatemapping.ca.

43 Spencer Van Dyk, "Canada Has Only Planted 29 Million of the 2 Billion Trees Promised by 2030," CTV News, July 8, 2022, ctvnews.ca.

44 Carlos Anchondo, "CCS 'Red Flag?': World's Sole Coal Project Hits Snag," *E&E News*, January 10, 2022, eenews.net.

45 Greg Muttitt, *Net Expectations: Assessing the Role of Carbon Dioxide Removal in Companies' Climate Plans* (Greenpeace UK, 2021), greenpeace.org.uk.

46 Justine Townsend, Faisal Moola, and Mary-Kate Craig, "Indigenous Peoples Are Critical to the Success of Nature-Based Solutions to Climate Change," *FACETS* 5, no. 1 (January 2020): 551, doi: 10.1139/facets-2019-0058.

47 "Indigenous Peoples currently steward around one-fifth of the total carbon sequestered by forests and Indigenous territories encompass 40% of protected areas in the world." *Decolonizing Climate Policy in Canada*, Indigenous Climate Action, 48. See also: Biofuelwatch, Climate Justice Alliance, Energy Justice Network, et al., *Hoodwinked in the Hothouse: Resist False Solutions to Climate Change* (2021), climatefalsesolutions.org;

Climate Alliance of European Cities with Indigenous Rainforest Peoples, *UNREDDY: A Critical Look at REDD+ and Indigenous Strategies for Comprehensive Forest Protection* (2015), climatealliance.org.

48    *Canada's Big Oil Reality Check: Major Oil and Gas Producers Undercut Canada's Commitment to 1.5°C* (Oil Change International & Environmental Defence Canada, 2021), 20, priceofoil.org.

49    For more, see: Kate Aronoff, *Overheated: How Capitalism Broke the Planet—and How We Fight Back* (Bold Type Books, 2021); and Harsha Walia, *Border and Rule: Global Migration, Capitalism, and the Rise of Racist Nationalism* (Halifax & Winnipeg: Fernwood, 2021).

## 3. A Just Fossil Fuel Phase-Out

1    While there were reports before and during the 2021 election that Liberals had a credible climate plan, this speculation was almost all based on one opaque and unreleased climate modelling exercise done by economist Mark Jaccard: Mark Jaccard, "Assessing Climate Sincerity in the Canadian 2021 Election," *Policy Options*, September 3, 2021, policyoptions.irpp.org. A line-by-line analysis of the 2021 Liberal climate platform by co-author David Gray-Donald did not find measures that would clearly and directly lead to a reduction in the production and consumption of fossil fuels in Canada: David Gray-Donald, "Analysis: The Liberal Climate Plan Is New Denialist Trash," Media Co-op, September 16, 2021, mediacoop.ca.

2    *Net Zero by 2050: A Roadmap for the Global Energy Sector; Summary for Policy Makers* (International Energy Agency, 2021), 11, iea.org.

3    A 2022 Tyndall Centre study found that for a 67 percent chance at limiting heating to 1.5°C, Canada and peers would need to phase out all production by 2030–32, and the last/poorest countries would need to follow suit by 2040, with support from the Global North. Even this timeline is longer than "fair" for Canada, given historic emissions, but it is the minimum that's societally plausible, and so the authors propose wealthy countries pursue equity by heeding calls for debt justice, climate finance, and other international reparations. See: Dan Calverley and Kevin Anderson, "Fossil Fuel Extraction Pathways within Paris-Compliant Budgets" (Tyndall Centre for Climate Change Research, March 2022).

4    Ploy Achakulwisut and Peter Erickson, *Trends in Fossil Fuel Extraction: Implications for a Shared Effort to Align Global Fossil Fuel Production with Climate Limits* (Stockholm: Stockholm Environment Institute, 2021), 15, doi: 10.51414/sei2021.001.

5    *Net Expectations: Assessing the Role of Carbon Dioxide Removal in Companies' Climate Plans* (Greenpeace UK, 2021), 8, greenpeace.org.

6    "New Tax Credit for Carbon Capture Will Only Delay Climate Action, Warns Environmental Advocate," CBC Radio, April 8, 2022, cbc.ca.

7    Amanda Stephenson, "Federal Tax Credit Not Enough to get Carbon
     Capture Projects Built: Cenovus CEO," CTV News, April 27, 2022,
     calgary.ctvnews.ca; Robert Tuttle, "Oil Sands Carbon Cuts Come
     with US$60-Billion Bill, Loose Ends," *Bloomberg News*, July 8, 2021,
     bnnbloomberg.ca.

8    Danielle Cameron, "Top Active Coal Mines in Canada," *Canadian Mining
     & Energy*, April 11, 2019, miningandenergy.ca.

9    Commissioner of the Environment and Sustainable Development to
     the Parliament of Canada, *Report 3—Hydrogen's Potential to Reduce
     Greenhouse Gas Emissions* (Office of the Auditor General of Canada, April
     26, 2022), oag-bvg.gc.ca.

10   Natasha Bulowski, "New Climate Plan's Reliance on Carbon Capture
     Called 'Not at All Realistic,'" *Canada's National Observer*, March 30, 2022,
     nationalobserver.com.

11   Mazan Labban, "Oil in Parallax: Scarcity, Markets, and the
     Financialization of Accumulation," *Geoforum* 41, no. 4 (July 2010), doi:
     10.1016/j.geoforum.2009.12.002.

12   Greg Muttitt et al., *The Sky's Limit: Why the Paris Climate Goals Require
     a Managed Decline of Fossil Fuel Production* (Oil Change International,
     2016), 18, priceofoil.org.

13   Fergus Green and Richard Denniss, "Cutting with Both Arms of the
     Scissors: The Economic and Political Case for Restrictive Supply-Side
     Climate Policies," *Climatic Change* 150 (March 2018), 73–87, doi: 10.1007/
     s10584-018-2162-x.

14   The Next System Project and Oil Change International, *The Case for
     Public Ownership of the Fossil Fuel Industry: Don't Bail Out Fossil Fuel
     Executive—Take Control for Workers & Communities* (April 2020),
     priceofoil.org.

15   These figures are the average for each year where data is available from
     2017 to 2019, or additional years if these are not available. Figures fluctuate
     year to year as they are tied to the price of oil or come from one-off sup-
     port programs. The International Institute for Sustainable Development
     (IISD) estimates that the average annual direct fossil fuel subsidies from
     federal and provincial governments were $4.8 billion per year in 2018 and
     2019. Government of Canada Parliamentary Budget Officer data shows
     an additional $2 billion on average through tax deductions from 2017 to
     2019. Canadians for Tax Fairness estimates $3 billion a year in carbon tax
     loopholes for large emitters that are largely fossil fuel companies (no year
     specified). Finally, Oil Change International analysis shows $11 billion a
     year in general public finance for fossil fuels through Export Development
     Canada plus $3 billion a year 2017 to 2019 for pipeline support through
     EDC (Trans Mountain Expansion and Coastal GasLink). See: Tara Laan

and Vanessa Corkal, *International Best Practices: Estimating Tax Subsidies for Fossil Fuels in Canada* (Winnipeg and Geneva: IISD, December 2020), iisd.org; Philip Bagnoli and Tim Scholz, *Energy Sector and Agriculture: Federal Revenue Forgone from Tax Provisions* (Parliamentary Budget Office, 2021), pbo-dpb.gc.ca; Bronwen Tucker and Kate DeAngelis, *Past Last Call: G20 Public Finance Institutions Are Still Bankrolling Fossil Fuels* (Oil Change International and Friends of the Earth US, October 2021), priceofoil.org.

16 "Wilkinson Says Aid Program to Cut Methane Emissions Will Be Reconsidered," Canadian Press, November 29, 2021, jwnenergy.com; Natasha Bulowski, "Critics Want Trudeau to Dump Oil and Gas Fund," *Canada's National Observer*, November 29, 2021, nationalobserver.com.

17 Oil Change International, "Public Finance for Energy Database— Download Dataset," accessed April 2022, energyfinance.org.

18 Many upstream (production only) oil and gas companies like Cenovus, which sell unrefined or less refined products, advocated for the temporary production cuts, as under the "oversupply" dominant in the market in 2018 they could only get very low prices from refineries for their crude oil. In comparison, integrated companies such as Suncor, which produce, refine, and sell finished products to consumers, generally opposed the production cuts because they could offset upstream losses with the benefits low prices gave to the refining portions of their businesses. The 2018 policy of production cuts exposed divisions within the industry. See: Josh Wingrove and Kevin Orland, "Oil Industry Spars over Forcing Temporary Production Cuts as Heavy Crude Dips to $15," *Bloomberg News*, November 15, 2018, financialpost.com.

19 Mike De Souza, Carolyn Jarvis, Emma McIntosh, and David Bruser, "Alberta Regulator Privately Estimates Oilpatch's Financial Liabilities Are Hundreds of Billions More Than What It Told the Public," *Canada's National Observer*, November 1, 2018, nationalobserver.com.

20 Regan Boychuk, Mark Anielski, John Snow Jr., and Brad Stelfox, *The Big Cleanup: How Enforcing the Polluter Pay Principle Can Unlock Alberta's Next Great Jobs Boom* (Alberta Liabilities Disclosure Project, June 2021), aldpcoalition.com.

21 Some of this industrial planning should be possible in existing federal ministries (as there are many more ministries now than there were during the war). The Ministry of Environment and Climate Change as well as the Ministry of Natural Resources could house some just transition activities, but that has not yet happened to nearly the extent required. There can be many reasons for ministry inaction, including the elected political leadership and also the permanent ministry staff bureaucrats, a number of whom are targets of prolonged lobbying by the oil and gas industry.

To the extent that creating a new ministry, agency, or Crown corporation would help avoid institutional inertia and initiate meaningful action, it is a strategy worth pursuing. For more, see: Seth Klein, *A Good War: Mobilizing Canada for the Climate Emergency* (ECW Press, 2020).

22  Ministry of Just Transition Collective, "The Year Is 2025, and a Just Transition Has Transformed Canada," *Breach*, March 18, 2022, breachmedia.ca.

23  Kevin Taft, *Oil's Deep State: How the Petroleum Industry Undermines Democracy and Stops Action on Global Warming—in Alberta, and in Ottawa* (Lorimer, 2017).

24  "Proxy Alert: Enbridge Inc.–Shareholder Proposal on Environmental and Indigenous Rights Due Diligence," Shareholder Association for Research & Education, April 12, 2017, share.ca; Erin Ellis, "Nuns and Standing Rock Leader Press Enbridge Shareholders for Change," *Canada's National Observer*, May 12, 2017, nationalobserver.com; Aaron Labaree, "NoDAPL: Standing Rock and the 'Deep North,'" Al Jazeera, January 17, 2017, aljazeera.com.

25  Starblanket, "Crises of Relationship," 20.

26  Hadrian Mertins-Kirkwood and Zaee Deshpande, *Who Is Included in a Just Transition?: Considering Social Equity in Canada's Shift to a Zero-Carbon Economy* (Canadian Centre for Policy Alternatives and Adapting Canadian Work and Workplaces to Respond to Climate Change, August 2019), 13–16, policyalternatives.ca.

27  Ian Hussey, *The Future of Alberta's Oil Sands Industry: More Production, Less Capital, Fewer Jobs* (Parkland Institute, March 2020), 11, parklandinstitute.ca.

28  Ian Hussey and Emma Jackson, *Alberta's Coal Phase-Out: A Just Transition?* (Parkland Institute, November 2019), 41, 46, parklandinstitute.ca.

29  David Gray-Donald and Emily Eaton, "A Just Transition Requires a Planned Economy: But Whose Plan?," *Briarpatch*, October 10, 2019, briarpatchmagazine.com.

## 4. Green Infrastructure for All

1  Vickie Wetchie, online interview with authors, May 21, 2021.

2  Alex Hemingway, "Wealth Tax Would Raise Far More Money Than Previously Thought," *Policy Note* (Canadian Centre for Policy Alternatives), March 11, 2021, policynote.ca.

3  Many reports and authors make the point that progressive taxation is a form of climate change policy in itself. The Cambridge Sustainability Commission Report on Scaling Behaviour Change reported in 2021 that "over the period 1990–2015, nearly half of the growth in absolute

global emissions was due to the richest 10%, with the wealthiest 5% alone contributing over a third (37%)." See: Peter Newell, Freddie Daley, and Michelle Twena, *Changing Our Ways?: Behaviour Change and the Climate Crisis* (Cambridge Sustainability Commission on Scaling Behaviour Change, 2021), rapidtransition.org.

4   Alex Hemingway, "A Modest Wealth Tax Would Raise $363B over 10 Years," *Breach*, September 14, 2021, breachmedia.ca.

5   Canadians for Tax Fairness, "Briefing Package: Fair Tax Priorities for Budget 2022," February 2022, taxfairness.ca.

6   Resource Movement, "Resource Movement Reiterates Its Call for a Federal Wealth Tax," November 5, 2020, resourcemovement.org.

7   See: Johanna Bozuwa, "Community Ownership of Power Administration: Putting Utilities under Public Control," The Next System Project, February 2019, thenextsystem.org; Martin Adelaar, Roger Peters, and Geoff Stiles, "Energy Democracy: An Essential Component of Social Democracy," Broadbent Institute, December 8, 2017, broadbentinstitute.ca.

8   Sean Sweeney, Kylie Benton-Connell, and Lara Skinner, "TUED Working Paper #4: Power to the People," Trade Unions for Energy Democracy, rosalux.nyc. In Navajo Nation, successful efforts from Diné C.A.R.E and other grassroots groups to close the massive Navajo Generating Station coal plant and Kayenta Mine and secure just transition funding have also led to a Light Up Navajo pilot program for distributed solar energy in households without energy access from the Navajo Tribal Utility Authority (NTUA). However, there is still a lot of work to do. As of December 2021, NTUA had added connections in just 737 of the 15,000 households without electricity access in the Nation, who together make up 75 percent of all unelectrified households in the United States. See: Navajo Tribal Utility Authority, "Light Up Navajo III—A Project to Fulfill the Hopes for Electric Connection Continues," ntua.com.

9   Justin Brake and Ashley Brandson, "Hydro Had 'Bigger Impact' Than Residential School in Misipawistik: Councillor," *APTN National News*, September 18, 2018, aptnnews.ca.

10  In his 2020 report into the financial and operational issues facing the project, Supreme Court Justice Richard LeBlanc suggested the Government of Newfoundland and Labrador and the Crown corporation Nalcor "created an environment of mistrust" and that even today the government and the company have failed to ensure that their commitments "regarding environmental matters related to the project are being properly tracked, monitored and addressed." Cited in: "Inquiry Finds Environment of 'Mistrust' after Lack of Indigenous Consultations for Muskrat Falls Project," *APTN National News*, March 13, 2020, aptnnews.ca.

11  This figure was calculated using two different tables (National + Economic Sector) from Environment and Climate Change Canada, "Greenhouse Gas Emissions," canada.ca.

12  Karine Godin, "Maisons a consommation energetique nette zero au Nouveau-Brunswick" (Université de Moncton, 2022).

13  See: Dana Cook "A Powerful Landscape: First Nations Small-Scale Renewable Energy Development in British Columbia" (University of Victoria, 2019).

14  *2030 Emissions Reduction Plan: Canada's Next Steps for Clean Air and a Strong Economy* (Environment and Climate Change Canada, 2022), canada.ca.

15  Farrah Merali, "Investors Now Make Up More Than 25% of Ontario Homebuyers, Pushing Prices Higher, Experts Warn," CBC News, November 23, 2021, cbc.ca; Martine August, "The Financialization of Canadian Multi-Family Rental Housing: From Trailer to Tower," *Journal of Urban Affairs* 42, no. 7 (2020), doi: 10.1080/07352166.2019.1705846.

16  John Clarke, "When Canada's Housing Bubble Pops, It Will Cause Misery and Ruin," *Jacobin*, January 6, 2022, jacobinmag.com.

17  Indigenous Services Canada, "Ending Long-Term Drinking Water Advisories," updated February 24, 2022, sac-isc.gc.ca.

18  Randy Shore, "Metro Vancouver First Nations 'Buying Their Own Land Back': Thousands More Housing Units Ready to Break Ground," *Vancouver Sun*, November 7, 2019, vancouversun.com.

19  Sogorea Te' Land Trust, sogoreate-landtrust.org; Deonna Anderson, "These Indigenous Women Are Reclaiming Stolen Land in the Bay Area," *Yes Magazine*, April 30, 2019, yesmagazine.org.

20  Shiri Pasternak, Naiomi Metallic, Yumi Numata, Anita Sekharan, Jasmyn Galley, and Samuel Wong, *Cash Back: A Yellowhead Institute Red Paper* (Yellowhead Institute, 2021), 58, yellowheadinstitute.org.

21  Pasternak et al., *Cash Back*, 61.

22  Kate Aronoff, Alyssa Battistoni, Daniel Aldana Cohen, and Thea Riofrancos, *A Planet to Win: Why We Need a Green New Deal* (Verso, 2019).

23  Stephenson Strobel, Ivana Burcul, Jia Hong Dai, Zechen Ma, Shaila Jamani, and Rahat Hossain, "Characterizing People Experiencing Homelessness and Trends in Homelessness Using Population-Level Emergency Department Visit Data in Ontario, Canada" (Statistics Canada, 2021), doi: 10.25318/82-003-x202100100002-eng.

24  "Understanding Core Housing Need," Canadian Mortgage and Housing Corporation, August 14, 2019, cmhc-schl.gc.ca.

25  *Energy Poverty in Canada: A CUSP Backgrounder* (Canadian Urban Sustainability Practitioners, October 2019), 2, energypoverty.ca.

26  Gregory Suttor, "Canadian Social Housing: Policy Evolution and Impacts on the Housing System and Urban Space" (PhD diss., University of Toronto, 2014), 106, utoronto.ca.

27  Suttor, "Canadian Social Housing," 112.

28  David Hulchanski, "The Invention of Homelessness," *Toronto Star*, September 18, 2010.

29  "Venezuela: Social Program Meets Goal, Delivers 3 Million Homes," TeleSur, December 27, 2019, telesurenglish.net.

30  According to the Environment and Climate Change Canada, "in 2019, the transport sector was the second largest source of GHG emissions, accounting for 25% (186 megatonnes of carbon dioxide equivalent) of total national emissions." *Greenhouse Gas Emissions: Canadian Environmental Sustainability Indicators* (Environment and Climate Change Canada, 2021), 9, canada.ca.

31  *Greenhouse Gas Emissions* (Environment and Climate Change Canada).

32  Zeke Hausfather, "Factcheck: How Electric Vehicles Help to Tackle Climate Change," Carbon Brief, May 13, 2019, carbonbrief.org.

33  "Highway of Tears: Preventing Violence Against Women," Carrier Sekani Family Services, highwayoftears.org.

34  Thomas Blampied, "End of the Line?: The History of Canada's Precarious Passenger Rail Network," Active History, February 21, 2020, activehistory.ca.

35  James Wilt, *Do Androids Dream of Electric Cars?: Public Transit in the Age of Google, Uber, and Elon Musk* (Toronto: Between the Lines, 2020), 10.

36  Wilt, *Do Androids Dream of Electric Cars?*, 3.

37  Emily Eaton and Simon Enoch, *Renewable Regina: Putting Equity into Action* (Regina: Canadian Centre for Policy Alternatives, 2020), policyalternatives.ca.

38  There are many good resources and research to consult on each of these solutions. One helpful website that gathers many resources and publishes its own summaries is the Sustainable Urban Transport Project at sutp.org.

39  "TTC Electrical Workers Call for Free Public Transit and a Campaign of Mass Strikes and Protests to Bring Down Doug Ford," Canadian Union of Public Employees Local 2, July 2019, cupelocal2.com.

40  ATU Canada and Emily Leedham, "Still Waiting for the Bus: The Unnatural Death of Prairie Intercity Transit," ATU Canada, October 2019, atucanada.ca.

## 5. miyo-wîcihtowin

1  Emma Jackson, "Caring for Crude in an Era of Capitalist Crisis: Migrant Caregivers and the Fort McMurray Wildfire" (MA diss., University of Alberta, 2019), 3–9, ualberta.ca.

2  Canadian Centre for Policy Alternatives, The Canadian Women's

Foundation, and Ontario Nonprofit Network, *Recovery through Equality: Developing an Inclusive Action Plan for Women in the Economy* (Canadian Centre for Policy Alternatives, December 2020), 2, policyalternatives.ca.

3 Winona LaDuke and Deborah Cowen, "Beyond Wiindigo Infrastructure," *South Atlantic Quarterly* 119, no. 2 (April 2020): 243–68, doi: 10.1215/00382876-8177747.

4 Pasternak et al., *Cash Back*, 7.

5 Pasternak et al., *Cash Back*, 29.

6 Pasternak et al., *Cash Back*, 29.

7 Anya Zoledziowski and Natashya Gutierrez, "Land Defenders Are Killed in the Philippines for Protesting Canadian Mining," *Vice*, October 1, 2020, vice.com.

8 Pasternak et al., *Cash Back*, 33.

9 LaDuke and Cowen, "Beyond Wiindigo Infrastructure," 244, cited in Pasternak et al., *Cash Back*, 7.

10 Kim Anderson, *A Recognition of Being: Reconstructing Native Womanhood* (Toronto: Sumach Press, 2000), 158–59.

11 Angele Alook, "Indigenous Families: Migration, Resistance, and Resilience," in *Continuity and Innovation: Canadian Families in the New Millennium*, ed. Amber Gazso and Karen Kobayashi (Toronto: Nelson Education, 2017), 106.

12 Alook, "Indigenous Families," 106.

13 Amelia M. Paget, *People of the Plains* (Regina: Canadian Plains Research Center, 2004), 39, cited in: Kim Anderson, *Life Stages and Native Women: Memory, Teachings, and Story Medicine* (Winnipeg: University of Manitoba Press, 2011), 66.

14 Anderson, *Life Stages and Native Women*, 73.

15 Anderson, *Life Stages and Native Women*, 65.

16 Sylvia McAdam (Saysewahum), *Nationhood Interrupted: Revitalizing nêhiyaw Legal Systems* (Saskatoon: Purich Publishing, 2015), 29, 36.

17 Johanne Johnson, "miyo-wîcihtowin bundle" (Wahkohtowin Law and Governance Lodge, University of Alberta, 2018), ualberta.ca.

18 Johanne Johnson, "sihtoskâtowin bundle" (Wahkohtowin Law and Governance Lodge, University of Alberta, 2018), ualberta.ca.

19 Johanne Johnson, "manâcihtâwin bundle" (Wahkohtowin Law and Governance Lodge, University of Alberta, 2018), ualberta.ca.

20 Pasternak et al., *Cash Back*, 67.

21 Alook, Hussey, and Hill, "Indigenous Gendered Experiences of Work."

22 "Decent Work," International Labour Organization, ilo.org.

23 Simpson, "The State Is a Man."

24 Pasternak et al., *Cash Back*, 8.

25 Pasternak et al., *Cash Back*, 9.

26  Diane Elson, "Plan F: Feminist Plan for a Caring and Sustainable
    Economy," *Globalizations* 13, no. 6 (2016), 919–21, doi:
    10.1080/14747731.2016.1156320.

27  The "proposed dental program would start with those under 12 years
    old in 2022, then expand to under-18-year-olds, seniors and persons
    living with a disability in 2023," and would be fully implemented by 2025.
    Catharine Tunney, "Liberals Agree to Launch Dental Care Program in
    Exchange for NDP Support," CBC News, March 22, 2022, cbc.ca.

28  The agreement calls for "a Canada Pharmacare Act passed by the end of
    2023 to task the National Drug Agency to develop a national formulary of
    essential medicines and a bulk purchasing plan by the end of the agree-
    ment." Lauren Pelley, "Why Pharmacare Plans Keep Stalling in Canada—
    Even as Research Suggests Billions in Savings," CBC News, April 6, 2022,
    cbc.ca.

29  Jim Stanford, "The Role of Early Learning and Childcare in Rebuilding
    Canada's Economy after COVID-19" (Centre for Future Work and
    the Canadian Centre for Policy Alternatives, November 2020); Kerry
    McCuaig and Tara McWhinney, "Policy Briefing Note: The Early
    Childhood Education and Workforce" (Canadian Research Institute for
    the Advancement for Women, 2017), criaw-icref.ca.

30  Jérôme De Henau and Susan Himmelweit argue for a recovery plan takes
    into consideration both physical infrastructure and building a better social
    infrastructure. In their comparative analysis of eight OECD countries
    examining investments in construction versus care economies (childcare
    and eldercare), they found that a "care-led, rather than construction
    recovery program creates more jobs and reduces gender equality" (456).
    While construction projects tend to create short-term jobs, investing
    in care systems improves society as a whole and creates long-term jobs
    and long-term benefits from gender equality in the labour market.
    Jérôme De Henau and Susan Himmelweit, "A Care-Led Recovery from
    Covid-19: Investing in High-Quality Care to Stimulate and Rebalance
    the Economy," *Feminist Economics* 27, no. 1 & 2, (2021): 453–69, doi:
    10.1080/13545701.2020.1845390.

31  Jim Selby and Garry Sran, "COVID-19 Pandemic Exposes Senior Care
    Failure: Now Is the Time to Build a Public Pan-Canadian Senior Care
    System" (Alberta Union of Provincial Employees, February 26, 2021), 13.

32  Selby and Sran, "COVID-19 Pandemic Exposes Senior Care Failure,"
    22–23.

33  George Monbiot, "Public Luxury for All or Private Luxury for Some: This
    Is the Choice We Face," *Guardian*, May 31, 2017, theguardian.com.

34  Canada's militarism is often ignored, but we have repeatedly followed
    more powerful nations like the United States and United Kingdom

into coups, oil wars, and other unethical foreign interventions, all while denying our culpability and instead casting ourselves as noble peacekeepers in the Canadian public imagination. Linda McQuaig summed this up memorably in the title of her book on Canadian imperialism *Holding the Bully's Coat* (Doubleday Canada, 2010). But outside of playing important roles in supporting other Western nations' imperialism, the Canadian military has often taken the lead in protecting Canadian corporate interests, such as mining operations abroad, to ensure their profits and resources continue to flow to these corporations and the Canadian state. See: Jeffery R. Webber and Todd Gordon, *Blood of Extraction: Canadian Imperialism in Latin America* (Halifax & Winnipeg: Fernwood, 2016). We also acknowledge that prisons, child welfare, policing, and border services are major employers of Indigenous, Black, and racialized folks, but if these institutions of death and colonial violence were abolished, and labour and resources redirected to hospitals, schools, childcare and eldercare centres, restorative justice programs, and other centres of healing, workers could transfer their skills to these jobs in economies of care. We don't want folks to lose their jobs, but to have access to decent, good jobs.

35   Robyn Maynard, *Building the World We Want: A Roadmap to Police Free Futures* (Building the World We Want Collective, 2021), 6, buildingtheworldwewant.com.

36   See the chapter "Green Anti-imperialism and the National Question," in Ajl, *A People's Green New Deal*.

### 6. Changing the Political Weather

1   Wealth tax and tax loopholes: Abacus Data, "Canadians Think Their Tax System Is Unfair and Support New Revenue Tools That Bring Down the Deficit and Reduce Inequality Now," August 4, 2021, abacusdata.ca; transition with financial support: Canadian Centre for Policy Alternatives, "Canadians Ready for Bold Policies to Transition away from Fossil Fuels, Poll Finds," August 12, 2019, policyalternatives.ca; TRC calls: Abacus Data for Assembly of First Nations, "Canadians React to the Discovery of Remains at Residential School," June 2021, afn.ca; police budget redirection: Andrew Russell, "Defund the Police? Canadians Split along Generational Lines, Ipsos Poll Suggests," Global News, July 25, 2020, globalnews.ca; Wet'suwet'en blockades: Angus Reid Institute, "Coastal Gaslink Chaos: Two-in-Five Support Protesters in Natural Gas Project Dispute; Half Support Pipeline," February 13, 2020, angusreid.org.

2   After the student strikes, Quebec post-secondary tuition was tied to an index of average disposable household income rather than frozen or made free as student unions demanded. Politicians who have proposed increases have all reneged on this position once in office. Marco Bélair-

Cirino, "Le Printemps érable, source d'avancées sociales et d'embarras
politique," *Le Devoir*, February 11, 2022, ledevoir.com.

3   Alain Savard, "Keeping the Student Strike Alive," *Jacobin*, September 2016,
jacobinmag.com; Richard Fidler, "Whither the Quebec Left and Student
Movement after the 'Maple Spring'?," *Canadian Dimension*, January 2013,
canadiandimension.com; Alessandro Drago, "Strikes, General Assemblies
and Institutional Insurgency: Explaining the Persistence of the Québec
Student Movement," *Social Movement Studies* 20, no. 6 (2021): 652–68,
doi: 10.1080/14742837.2020.1837100.

4   Jenny Stanley, *The Legacy of Occupy 10 Years Later*, podcast, Rosa
Luxemburg Stiftung New York Office, rosalux.nyc.

5   Mark Engler and Paul Engler, "André Gorz and the Path between Reform
and Revolution," *This Is an Uprising*, October 2021, thisisanuprising.org.

6   Dean Spade, "Solidarity Not Charity: Mutual Aid for Mobilization and
Survival," *Social Text* 38, no. 1 (2020): 131–51, deanspade.net.

7   Emilia Belliveau, James K. Rowe, and Jessica Dempsey, "Fossil Fuel
Divestment, Non-reformist Reforms, and Anti-capitalist Strategy," in
*Regime of Obstruction*, ed. Carroll, 463. The Stop the Money Pipeline
coalition in the United States (stopthemoneypipeline.com) is one
example of private bank campaigning that has leveraged its momentum
at single institutions towards financial regulatory reforms that are likely
to ultimately weaken US banks' overall power as well as their ability to
invest in fossil fuels (though these are still in development at the time of
writing).

8   There is ongoing debate about the degree to which private equity will be
able to facilitate significant fossil fuel expansion plans. Oil, gas, and coal
extraction in new regions usually need new pipelines, export terminals,
and other infrastructure to get to market. These megaprojects have historically required many large financial institutions—usually including some
government-backed public finance or insurance—to weather government
approval processes and to raise enough capital. It is safe to say that if a
sizable portion of major banks meaningfully exclude fossil fuel investments, then fewer fossil fuel projects will get built; it is less certain how
many fewer. See, for example: Kate Aronoff, "Private Equity Is Quietly
Keeping Fossil Fuel Companies in Business," *New Republic*, October 2021,
newrepublic.com.

9   Donald Gutstein, *Fossilized Finance: How Canada's Banks Enable Oil and
Gas Production* (Canadian Centre for Policy Alternatives, April 2021),
policyalternatives.ca.

10  Hayden King and Riley Yesno, "Part Four: Reclamation," in *Land Back*.

11  King, Pasternak, and Yesno, *Land Back*, 58.

12  RAVEN Trust, raventrust.com.

13  "Stranded Baffinland Mine Workers Pen Open Letter to Protesters, Say
     They Support Inuit," *CBC News*, February 11, 2021, cbc.ca.

14  Jessica Clogg, Hannah Askew, Eugene Kung, and Gavin Smith, *Indigenous
     Legal Traditions and the Future of Environmental Governance in Canada*
     (West Coast Environmental Law, September 2016), wcel.org.

15  US-based groups Movement for Black Lives (m4bl.org) and The Red
     Nation (therednation.org) both use a "Divest and Reinvest" framework to
     help make their overall visions for societal transformation more tangible.
     And Toronto-based Black feminist writer, activist, and educator Robyn
     Maynard (robynmaynard.com) frequently speaks about defunding as
     worldbuilding.

16  *The Demand Is Still #DefundThePolice: Lessons from 2020*, Interrupting
     Criminalization, interruptingcriminalization.com.

17  Jocelyn Timperley, "Why Fossil Fuel Subsidies Are So Hard to Kill,"
     *Nature News*, October 2021, nature.com.

18  Seth Klein, "Biden Beating Trudeau on Climate Leadership," *National
     Observer*, April 16, 2021, nationalobserver.com.

19  This figure includes the RCMP but does not include railway and military
     police or government departments enforcing specific statutes, i.e., income
     tax, customs and excise, immigration, fisheries and wildlife. Patricia
     Conor et al., "Police Resources in Canada, 2019" Statistics Canada,
     statcan.gc.ca.

20  Government of Canada, "Infographic for Canada Border Services
     Agency," Infobase, tbs-sct.gc.ca; "Financial Facts on Canadian Prisons,"
     John Howard Society of Canada, August 2018, johnhoward.ca.

21  Robin Maynard, *Policing Black Lives: State Violence in Canada from Slavery
     to the Present* (Halifax & Winnipeg: Fernwood, 2017).

22  "Resources," Building the World We Want Collective,
     buildingtheworldwewant.squarespace.com.

23  Department of National Defence, "March 2020—Defence Budget,"
     canada.ca.

24  "No New Fighter Jets for Canada," Canadian Foreign Policy Institute,
     December 2021, foreignpolicy.ca.

25  Max Ajl, "Green Anti-Imperialism and the National Question," in *A
     People's Green New Deal*.

26  These figures are the average for each year where data is available from
     2017 to 2019, or additional years if these are not available. Figures fluctuate
     year to year as they are tied to the price of oil or come from one-off sup-
     port programs. The International Institute for Sustainable Development
     (IISD) estimates that the average annual direct fossil fuel subsidies from
     federal and provincial governments were $4.8 billion per year in 2018 and
     2019. Government of Canada Parliamentary Budget Officer data shows

an additional $2 billion on average through tax deductions from 2017 to 2019. Canadians for Tax Fairness estimates $3 billion a year in carbon tax loopholes for large emitters that are largely fossil fuel companies (no year specified). Finally, Oil Change International analysis shows $11 billion a year in general public finance for fossil fuels through Export Development Canada plus $3 billion a year 2017 to 2019 for pipeline support through EDC (Trans Mountain Expansion and Coastal GasLink). See: Laan and Corkal, *International Best Practices*; Bagnoli and Scholz, *Energy Sector and Agriculture*; Tucker and DeAngelis, *Past Last Call*.

27   The Canadian Centre for Policy Alternatives estimates conservatively that a wealth tax with rates of 1 percent on net worth over $20 million, 2 percent over $50 million, and 3 percent over $100 million would raise $36 billion a year on average over a decade. Canadians for Tax Fairness estimates we could raise an additional $62 billion a year through other tax reforms: $14 billion a year from adding measures to slow international tax avoidance and evasion, $3.5 billion a year from inheritance and luxury goods taxes, $13.2 billion a year from reversing cuts to the corporate tax rate, $7 billion a year with a financial activities tax, and $24 billion a year from increased taxes on investment income including capital gains, stock options, and corporate dividends. We have omitted fossil fuel subsidy measures and a one-time pandemic profits tax from their tally. See: Hemingway, "A Modest Wealth Tax"; Canadians for Tax Fairness, "Fair Tax Priorities for Budget 2022."

28   Researchers at Energy Policy Tracker found $12.5 billion a year in provincial and federal highway expansion projects in Canada in 2020 and 2021, not including projects earmarked for safety improvements or repairs. For aviation, we conservatively use $500 million a year based on 2017 to 2019 public finance to Bombardier. There were further aviation bailouts to Air Canada and others during COVID-19, but we did not count these as they were specific to the pandemic and may not continue at this level. "Canada," Energy Policy Tracker, energypolicytracker.org; Stephanie Levitz, "Justin Trudeau Hammered over Bombardier Bailout in House of Commons," Global News, April 3, 2017, globalnews.ca.

29   Workers at the Giant Yellowknife Gold Mine, concerned about their own contact with arsenic, allied with members of the Dene Nation when they found out the pipeline with safe drinking water provided to settler communities had not been extended to the Nation. Their union, USWA, and the National Indian Brotherhood (now Assembly of First Nations) conducted a joint study to document the health implications of this arsenic exposure. Their collaboration ultimately led to new arsenic treatment infrastructure and federal regulations that greatly reduced (though did not entirely remove) the threat to both the Dene Nation and mine

workers. Katrin MacPhee, "Canadian Working-Class Environmentalism, 1965–1985," *Labour / Le Travail* 74 (Fall 2014), muse.jhu.edu.

30  In theory, construction union LiUNA has such an agreement, but in practice it is rather toothless as it only recognizes the federal government's assessments of Indigenous rights. See: Justine Hunter, "Construction Union, Assembly of First Nations Sign Pact to Promote Indigenous Workforce," *Globe and Mail*, June 11, 2017, theglobeandmail.com.

31  The last major general strike across Canada was in 1919, and there have been murmurs across the labour movement that another general strike is needed now. A general strike would likely carry some weight at this time, as the rich make record profits, minimum wages remain stagnant, and public sector workers and other essential workers have been unable to bargain for wage increases to match the unprecedented increases in inflation and increases in the cost of housing during the pandemic. Rachelle Friesen, "From Pandemic to Uprising: Lessons from the Winnipeg General Strike," *Spring*, August 26, 2021, springmag.ca.

32  Liz Walker and Shanice Regis-Wilkins, "What Is Sectoral Bargaining and How Can It Help Canada's Working Class?," Press Progress, July 30, 2021, pressprogress.ca.

33  Sharon Block and Benjamin Sachs, *Clean Slate for Worker Power: Building a Just Economy and Democracy* (Labor and Worklife Program, Harvard Law School, 2020).

34  Michal Rozworski, "Working Class Disarmed, Canadian Redux," blog, April 6, 2014, rozworski.org. Data: CANSIM, FRED, BLS. See also Statistics Canada, "Work Stoppages in Canada, by Jurisdiction and Industry," table 14-10-0352-01, statcan.gc.ca.

35  Under the federal Employment Equity Act, four groups are identified as equity-seeking groups: women, Indigenous peoples, persons with disabilities, and visible minorities (i.e., racialized people). But more broadly under human rights legislation, workers have the right to be free from discrimination under protected grounds in Canada, for example, in the Ontario Human Rights Code there are seventeen protected grounds: citizenship, race, place of origin, ethnic origin, colour, ancestry, disability, age, creed, sex/pregnancy, family status, marital status, sexual orientation, gender identity, gender expression, receipt of public assistance (in housing), and record of offences (in employment). In the labour movement, equity-seeking groups can be those protected under employment equity or human rights codes.

36  Fathima Cader, "Gig Workers Win the Right to Unionize," Canadian Centre for Policy Alternatives, May 1, 2020, policyalternatives.ca.

37  Karl Evers-Hillstrom, "More Than 50 Starbucks Stores Now Petitioning to Unionize," *Hill*, January 31, 2022, thehill.com.

38  Daniel Sarah Karasik, "Pandemic Work Refusals, and How to Build Capacity for More," Media Co-op, June 12, 2021, mediacoop.ca.

39  Kate Aronoff, "Striking Teachers in Coal and Gas Country Are Forcing States to Rethink Energy Company Giveaways," *Intercept*, April 12, 2018, theintercept.com.

40  Green Collective Agreements Database, yorku.ca.

41  Shane Gunster, "Extractive Populism and the Future of Canada," Canadian Centre for Policy Alternatives, July 2, 2019, policyalternatives.ca.

42  Emma Jackson, "What the Left Can Learn from the 'Freedom Convoy,'" *Breach*, February 2, 2022, breachmedia.ca; "The 'Freedom Convoy' Is Nothing but a Vehicle for the Far Right," Canadian Anti-Hate Network, January 27, 2022, antihate.ca.

43  Liz Walker and Shanice Regis-Wilkins, "Why Unionizing Canadian Workplaces Is More Important Now Than Ever during the Pandemic," Press Progress, March 19, 2021, pressprogress.ca.

44  Worker committees including human rights, environmental, young workers, BIPOC workers, anti-privatization, just transition, and climate crisis committees could all be relevant for this work. For more on caucuses, see: Ben Sichel, "Building a Caucus to Rebuild Union Power," *Rank and File*, January 21, 2022, rankandfile.ca.

45  The Transnational Institute has documented many other international examples of the "remunicipalisation" of public goods and services in their book *The Future Is Public*, ed. Satoko Kishimoto, Lavinia Steinfort, and Olivier Petitjean (Transnational Institute, 2020), tni.org.

## 7. Sihtoskâtowin

1  Pamela Palmater, "Resisting the Resurgence of White Supremacy," in *We Resist: Defending the Common Good in Hostile Times*, ed. Cynthia Levine-Rasky and Lisa Kowalchuk (McGill-Queen's University Press, 2020), 280–82.

2  "'Freedom Convoy' Is Nothing but a Vehicle," Canadian Anti-Hate Network.

3  Wayne Roberts and George Ehring, *Giving Away a Miracle: Lost Dreams, Broken Promises and the Ontario NDP* (Oakville, ON: Mosaic Press, 1993).

4  Robers and Ehring, *Giving Away a Miracle*, 52–54.

5  Jim Stanford, "The History of the New Politics Initiative: Movement and Party, Then and Now," November 29, 2011, rabble.ca.

6  David McGrane, *The New NDP: Moderation, Modernization, and Political Marketing* (Vancouver: UBC Press, 2019).

7  David Camfield, "The NDP and the Priorities of Radicals," *New Socialist*, June 30, 2017, newsocialist.org. Emphasis added.

8  Dru Oja Jay, "Movements Can Seize the Opportunities of a Second

Minority Government—Here's How," *Breach*, September 24, 2021, breachmedia.ca.

9   S.K. Hussan, "You Can't Change the World Alone, but All of Us Can Together," *Medium*, September 18, 2016, medium.com.

10  Justin Kong, Edward Hon-Sing Wong, and Veronica Yeung, "Organizing the Suburbs," *Briarpatch*, October 28, 2018, briarpatchmagazine.com.

11  The Red Nation, *The Red Deal: Indigenous Action to Save Our Earth* (Common Notions, 2021).

12  Emily Riddle, "mâmawiwikowin: Shared First Nations and Métis Jurisdiction on the Prairies," *Briarpatch*, September 2020, briarpatchmagazine.com.

13  Julia Travers, "The Movement for Black Lives Is on the Rise—and Funders Are Paying Attention," *Inside Philanthropy*, July 9, 2020, insidephilanthropy.com.

14  Yotam Marom, "Moving Toward Conflict for the Sake of Good Strategy," *Medium*, January 13, 2020, medium.com.

15  Maya Menezes, online interview with authors, June 25, 2021.

16  Ejeris Dixon, "Our Relationships Keep Us Alive: Let's Prioritize Them in 2018," *Truthout*, February 8, 2018, truthout.org.

17  Dr. Lilla Watson, a Murri elder and organizer, said this quote in a speech at the 1985 United Nations Decade for Women Conference in Nairobi. However, she has explained that the phrase came from collective work on Aboriginal rights in so-called Australia in the early 1970s and for this reason, she is not comfortable being identified as the sole author. See: Mz. Many Names, "Attributing Words," *Unnecessary Evils*, November 3, 2008, unnecessaryevils.blogspot.com.

18  Ricardo Levins Morales, "Whites Fighting Racism: What It's About," *Ricardo Levins Morales Art Studio*, January 7, 2015, rlmartstudio.wordpress.com.

19  Ross Andersen, "Protests across Canada Express Support for India Farmers," *CTV News*, December 5, 2020, ctvnews.ca.

20  Melissa Keith, "Educate, Agitate, Organize: The Halifax Workers' Action Centre," *Our Times*, October 9, 2019, ourtimes.ca. See also: Will Langford, "Friendship Centres in Canada, 1959–1977," *American Indian Quarterly* 40, no. 1 (2016): 1, doi: 10.5250/amerindiquar.40.1.0001.

21  Mark Fisher, *Capitalist Realism: Is There No Alternative?* (Winchester, UK: Zero Books, 2010).

22  Common Wealth: A Think-Tank for the Future, common-wealth.co.uk.

23  Sandy Garossino, "A Data-Based Dismantling of Jason Kenney's Foreign-Funding Conspiracy Theory," *Canada's National Observer*, October 3, 2019, nationalobserver.com.

24  Linda Besner, "Why a Group of Rich Millennials Are Giving Their Money Away," *Maclean's*, December 10, 2019, macleans.ca.

25 Groundswell Community Justice Trust Fund, "Putting Your Money Where Your Movement Is," groundswellfund.ca.

26 Angela Roque, Anishnabek Nation Treaty Authority, cited on home page of Treaty Land Sharing Network, treatylandsharingnetwork.ca.

27 Many not-for-profits have been dependent on issue-based grants that make them beholden to wealthy donors instead of social movements and the public interest. While one solution to this is building stronger alternative institutions fully separate from these funders, another strategy is to unionize workers at not-for-profits to shift the base of power in these organizations. Collectively, multiple unionized not-for-profits could push foundations towards less harmful granting practices. See: INCITE! Women of Color Against Violence, *The Revolution Will Not Be Funded: Beyond the Non-Profit Industrial Complex* (Duke University Press, 2007).

28 These ideas are covered at length by climate justice writer Mary Annaïse Heglar in her essay "Climate Change Isn't the First Existential Threat," Zora, February 18, 2019, zora.medium.com; by Nick Estes in *Our History Is the Future: Standing Rock versus the Dakota Access Pipeline, and the Long Tradition of Indigenous Resistance* (Verso Books, 2019); and by Billy-Ray Belcourt in his memoir *A History of My Brief Body* (Two Dollar Radio, 2020).

# INDEX

and governments strategies, 3;
and green infrastructure, 88;
Indigenous governments, 75; and
just transition, 9 (*see also* just transition); and radical reformations,
157–8; vs. "self-government," 44;
and shareholder defeats, 76; and
SWAT, 13; and systemic racism,
28; treated as afterthought, 6;
and Treaties, 22–3, 23–4, 35–6;
Tŝilhqot'in Nation, 151; UNDRIP
as regulatory first step, 74–5. See
*also* land rights
individuals, 68–9, 89–90, 94–6
industrialization, 38
industry and environment, 46–50,
53–6, 69–70. See *also* climate
denialism/delayism; emissions;
*specific industries*
industry and land rights, 42–3, 44–5.
See *also specific industries*
inherent rights, 21, 23–4. See *also* land
rights
injunctions, 45–6
Intergovernmental Panel on Climate
Change (IPCC), 52–3
International Energy Agency, 63–4
international public finance, 72
Inuit, 40, 43
Inutiq, Kunnuk (Sandra), 43
Iron & Earth, 81

James Bay and Northern Quebec
Agreement, 43
James Bay Cree, 94
Jameson, Fredric, 172
jobs: decent work, 116, 124–5; fossil
fuel transition, 10; green jobs,
73, 80, 148; job losses, 30; public
luxuries, 124–5; as trading time for
money, 118–19. See *also* workers
Just Powers, 33

just transition: overview, 153; in
Canada, 8–9; decent work, 116,
124–5; definition of, 3; ending
death economy, 125–6; funding,
139–40; and Global South, 8,
11; imagining, 129–31 (*see also*
social movements); Indigenous
Environmental Network
definition, 36; principles for, 9–11;
public ownership, 74, 76–7, 91;
regulations, 74–6; regulatory first
steps, 71–4; shared jurisdiction
and dual governance, 77–9, 85,
94, 96–7, 102; supply vs. demand
side solutions, 69–70; as term,
7–8; workers, 80–3. See *also* care
economies; green infrastructure;
Indigenous laws; political alignment; social systems of care
Justice for Workers! 147

Kayenta Mine, 195n8
Keep Transit Moving, 151
Kenney, Jason, 23
Keystone XL pipeline, 54–5
Kinder Morgan, 54
kitimâkis, 119–20
Klein, Ralph, 51
Klein, Seth, 52, 74
Kyoto Protocol, 48

Labban, Mazen, 69
Laboucan-Massimo, Melina, 30–1
labour movement. See unions
LaDuke, Winona, 109, 111
Laforest, Joël, 4–5, 159
Lameman, Crystal, 5, 15–17, 117–21
Lameman, Ron, 114
Lancaster Stands Up, 169
land, set aside, 72
land and wellness, 16–17
Land Back, 34–5, 117, 136–9, 148–9